Biomechanics

FOR DUMMIES®

A Wiley Brand

Biomechanics
FOR
DUMMIES®
A Wiley Brand

by Steven T. McCaw, PhD

FOR
DUMMIES®
A Wiley Brand

Biomechanics For Dummies®

Published by: **John Wiley & Sons, Inc.,** 111 River Street, Hoboken, NJ 07030-5774, www.wiley.com

Copyright © 2014 by John Wiley & Sons, Inc., Hoboken, New Jersey

Published simultaneously in Canada

For general information on our other products and services, please contact our Customer Care Department within the U.S. at 877-762-2974, outside the U.S. at 317-572-3993, or fax 317-572-4002. For technical support, please visit www.wiley.com/techsupport.

Wiley publishes in a variety of print and electronic formats and by print-on-demand. Some material included with standard print versions of this book may not be included in e-books or in print-on-demand. If this book refers to media such as a CD or DVD that is not included in the version you purchased, you may download this material at http://booksupport.wiley.com. For more information about Wiley products, visit www.wiley.com.

Library of Congress Control Number: 2013958288

ISBN 978-1-118-67469-7 (pbk); ISBN 978-1-118-67475-8 (ebk); ISBN 978-1-118-67476-5 (ebk); ISBN 978-1-118-67479-6 (ebk)

Manufactured in the United States of America

10 9 8 7 6 5 4

Contents at a Glance

Table of Contents

Chapter 21: Ten Things You May Not Know about Biomechanics 347

Chapter 22: Ten Ways to Succeed in Your Biomechanics Course 355

Introduction

Students enrolled in a biomechanics class usually find themselves looking at movement in a whole new way. After years of participation in organized or recreational sports, dance, and fitness, they've become pretty good "movers" in at least one activity. And because they like to, they just get out and move. For many, the enjoyment of moving has piqued an interest in pursuing a career as a teacher, coach, fitness specialist, or clinical therapist to help others become better movers.

Biomechanics is often your first exposure to the science of explaining how and why things move. Mechanics is the science concerned with forces acting on objects, and mechanics by itself is a demanding topic. Biomechanics goes a step further by applying the principles of mechanics to a living body. And the human body is the most complex thing around.

Fortunately, all movement and injury share the same basic principles. The overriding principle of biomechanics is that force causes all movements and underlies all injury or training that occurs. Knowing the effect of force — in fact, just knowing more about force as the source of all movement — provides a solid foundation for knowing more about movement and injury. And that's what biomechanics is all about.

About This Book

No biomechanics book can show you how to apply biomechanics to every possible form of human movement — there are just too many ways the human body can move. But what a biomechanics book can show you is that biomechanics applies to every possible form of human movement.

Biomechanics For Dummies is a reference book on the "what" and "how" of biomechanics. The "what" relates to the explanation of the terminology and principles of biomechanics, and the "how" relates to solving all the pesky equations that pop up in any science.

I try to be as informal as possible in a book that explores the science of something as complex as movement of the human body. I also try to make it clear why you need to know the concepts and equations of mechanics. I stay logical and factually precise, while simplifying some extremely challenging ideas.

In the examples, I demonstrate and reinforce a step-by-step format to problem solving. A systematic approach to equation and problem solving is important for anyone working at any level in biomechanics, maybe even more so at the beginning.

Anything marked by a Technical Stuff icon provides a more in-depth discussion of whatever material is being explained where the icon appears. This is useful but not necessary information. Also, the text in sidebars (shaded gray boxes) provides more details about a topic. It's interesting information (or I wouldn't have included it), but you won't miss any explanation about a topic by skipping right over a sidebar.

Finally, within this book, you may note that some web addresses break across two lines of text. If you're reading this book in print and want to visit one of these web pages, simply key in the web address exactly as it's noted in the text, pretending as though the line break doesn't exist. If you're reading this as an e-book, you've got it easy — just click the web address to be taken directly to the web page.

Foolish Assumptions

While writing this book, I made the following assumptions about who would read it:

- ✔ You may be an undergraduate college student taking an introductory biomechanics course. Most likely, you're enrolled in a program in kinesiology, exercise science, physical education, or athletic training. You've probably heard that biomechanics is a tough course, with a lot of math and physics, and you're pretty nervous about how well you'll do in the class.

- ✔ You may be a parent or coach in a youth league, and you've heard people talking about "the biomechanics of <insert the activity of your choice here>." You want to know more about the topic because it seems interesting and useful.

- ✔ Your basic math skills are still in your head somewhere, but you don't necessarily like to use them all that much unless you have to. You consider trigonometry something best avoided, although you may remember the Pythagorean theorem and what a hypotenuse is.

- ✔ You may or may not have had a previous course in anatomy and physiology. And if you did, you learned the names, origins, and insertions of a lot of muscles. But you don't recall a lot of details about exactly how we use our muscles to move.

Basically, except for the second item in this list, I'm assuming you're me back when I took an introductory biomechanics class.

Icons Used in This Book

Icons are the little pictures you see sprinkled throughout the margins of this book. They draw your attention to key types of information:

The Tip icon highlights a quick summary of an idea, application, or definition, or gives you insight on using a shortcut step with an equation.

The Remember icon jogs your memory to facts and ideas touched on earlier in the chapter or book that are relevant to the section you're reading.

The Warning icon tells you're in a misstep zone — a theory or equation where I've commonly seen (or made) bad moves. Take extra care.

The Technical Stuff icon marks some additional information on the topic that's too good not to include, but not essential to understanding the concept, equation, or idea.

Beyond the Book

In addition to the material in the print or e-book you're reading right now, this product also comes with some access-anywhere goodies on the web. Check out the free Cheat Sheet at www.dummies.com/cheatsheet/biomechanics for information on how running shoes work, what causes low-back pain, and more.

In addition, I've written several articles on topics ranging from how to turn around in outer space to composing a resultant force vector from multiple vectors. They're available at www.dummies.com/extras/biomechanics.

Where to Go from Here

You can use *Biomechanics For Dummies* as a supplement to a course you're taking or on its own as a text to understand the basic principles of biomechanics. I've written each chapter to stand on its own, so you don't need to move through the book in order, beginning with Chapter 1.

As you read the book, you'll notice that the principles of biomechanics are intertwined. The material in some chapters builds off topics explained in depth in other chapters. For example, the explanation of force, covered in Chapter 4, is useful to understanding torque, explained in Chapter 8. As a convenience, I include cross-references in each chapter to guide you to more in-depth discussions.

If you're new to the subject of biomechanics, beginning with Chapter 1 is a good idea. If you're taking an introductory course in biomechanics, your instructor and/or your textbook may present the material in a different order than I use here. If so, you can jump around, supplementing your instructor's lectures and your textbook by reading the chapters of this book that are relevant to the material being covered in your course.

Part I
Getting Started with Biomechanics

getting started with

biomechanics

web extras

For Dummies can help you get started with lots of subjects. Visit www.dummies.com to learn more and do more with *For Dummies*.

In this part...

- ✔ Identify the "bio" and the "mechanics" parts of biomechanics.

- ✔ Get a refresher on the basic math and geometry skills you need to solve biomechanics problems.

- ✔ Discover a systematic approach to resolving or composing vectors using SOH CAH TOA.

- ✔ Understand the fundamental terms and concepts of biomechanics.

Chapter 1

Jumping Into Biomechanics

In This Chapter

▶ Defining biomechanics

▶ Introducing linear and angular mechanics

▶ Using biomechanics to analyze movement

Kinesiology is the science focused on the study of motion. It's the core area of many majors at colleges and universities for students interested in exercise or movement science, athletic training, and physical education teacher education. A degree in kinesiology can lead to a career in itself in teaching, exercise prescription, sports medicine, and coaching. In addition, many students study kinesiology at the undergraduate level because its focus on the human body provides a strong foundation for graduate study in physical therapy and medicine.

Biomechanics is one of the core courses in kinesiology. Along with the foundation knowledge from other core courses (including anatomy and physiology, psychology of sport and exercise, exercise physiology, and motor learning), biomechanics contributes to a basic understanding of human movement possibilities.

In this chapter, I introduce you to the subject of this book — think of this as the book in a nutshell — with plenty of cross-references so you know where to turn to find more information.

Analyzing Movement with Biomechanics

Biomechanics uses three branches of mechanics, along with the structure and function of the living body, to explain how and why bodies move as they do (see Figure 1-1). The different branches of mechanics are used to study movement in specific situations, and the systems of the living body determine what it's capable of doing and how it responds during movement.

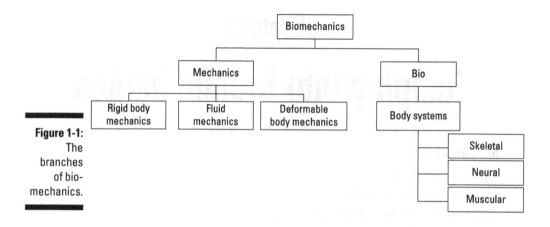

Figure 1-1:
The
branches
of bio-
mechanics.

In this section, I give you a brief overview of the three branches of mechanics, along with the structure and function of the living body.

Mechanics

Mechanics is a long-established field of study in the area of physics. It focuses on the effect of forces acting on a body. A *force* is basically a push or a pull applied to a body that wants to make it move (see Chapter 4). Mechanics looks at how a body is affected by forces applied by muscle, gravity, and contact with other bodies.

I use the term *body* to refer to the focus of attention during an analysis. For someone walking, the body could be the person as a whole entity. But the body could also be an individual segment, like the walker's thigh, lower leg, or foot, or, going even further, an individual bone in a segment. For more on defining the body under analysis, turn to Chapter 4.

Rigid body mechanics

An applied force affects the motion of a body — meaning, it tries to make the body speed up or slow down. The motion can be large and involve a lot of body segments, like walking, or it can be small and involve only a couple of segments, like bending a finger. Both of these movements, and all other movements involving body segments, can be analyzed using rigid body mechanics.

Rigid body mechanics simplifies a body by *modeling* (representing) it as a single, rigid bar. A rigid bar can be used to represent the entire body (quite a simplification) or just the individual segments of the body. The modeled segments can be combined as rigid, non-deforming links joined at hinges (the joints) to represent any part of the body.

Consider your arm, made up of the complex anatomical structures of the upper arm, forearm, and hand. If you hold your arm out in front of you and bend and straighten your wrist and elbow, you'll notice that your skin shifts and folds and soft areas bulge as muscles change shape under the skin. Place a finger over the front of your upper arm, and feel the changing stiffness of the muscle when your arm bends and straightens. If you poke your skin with a finger, it sinks in a little bit. In rigid body mechanics, these changes in, or deformations of, the individual segments are ignored. The upper arm, forearm, and hand are considered to be separate, simple rigid links or sticks that move at the joints where they meet. The rigid link model of the human body is more fully explained in Chapter 8.

Fluid mechanics

Fluid mechanics is the branch of mechanics focused on the forces applied to a body moving in air or water. These fluids produce forces called *lift* and *drag,* which affect the motion of a body when a fluid moves over it, or as the body moves through a fluid.

Fluid mechanics is obviously applicable to swimming and water sports, but it's also useful when explaining how to make a soccer ball, tennis ball, or baseball curve through the air. For more on fluid mechanics, float on over to Chapter 11.

Deformable body mechanics

Deformable body mechanics focuses on the changes in the shape of the body that are ignored in rigid body mechanics. An applied force causes a *deformation* (change in shape) of the body by loading the particles of material making up the body. Deformable body mechanics involves looking at the loading and the motion of the material within the body itself.

The loading applied to a body is called a *stress.* The size and the direction of the stress cause deformations of the material within the body, called *strain.* The relationship between the applied stress and the resulting strain is useful to understand injury to and training of tissues within the body. Chapter 12 provides more detail on deformable body mechanics.

Bio

Bio is Greek for "life," making biomechanics the science applying the principles of mechanics to a living body. Biomechanics is used to study and explain how and why living things move as they do, including the flight of a bumblebee, the swaying of a stalk of corn, and, more important for most of us, the movements of human beings.

Part IV of this book covers the "bio" of biomechanics, explaining aspects of the following systems important to the mechanics of movement:

✔ **Skeletal system:** The skeletal system, including bones, ligaments, and joints, provides the physical structure of the body and allows for movement. (See Chapter 13.)

✔ **Neural system:** The neural system, also known as the nervous system, including different types of nerve cells, serves as the communication system to control and respond to movement. (See Chapter 14.)

✔ **Muscular system:** The muscular system, including muscle and the tendon attaching muscle to bone, provides the motors we control to make our segments, and our bodies, move. (See Chapter 15.)

Later in this book, I give you an overview of the anatomy and function of the components of each of these systems and explain how each system influences movement.

Expanding on Mechanics

In mechanics, we look at how an applied force affects the motion of a body. Each branch of mechanics includes two subdivisions, one focused on describing the motion (kinematics) and the other focused on the forces that cause motion (kinetics). Figure 1-2 gives you a handy diagram of these subdivisions of mechanics, which I describe in more detail in this section.

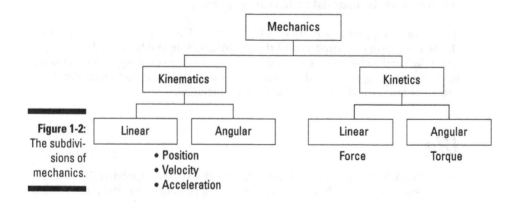

Figure 1-2:
The subdivisions of mechanics.

Describing motion with kinematics

Kinematics is the subdivision of mechanics focused on the description of motion. Kinematics is what we see happen to the body. When you watch a body, and describe its position, how far it travels, how fast it travels, and whether it's speeding up or slowing down, you're conducting a kinematic analysis.

Human movement is complex, even with simple moves. Try this: Use the tip of your index finger to draw a straight line across this page or screen. Can you do it if just your index finger moves? No, you get a short curved line. If just your hand moves at the wrist? No, you get a long, but still curved, line. If just your forearm moves at the elbow? No, you get a longer curved line. To make the tip of your finger move in a straight line across the page, you must coordinate the movement of at least two joints: the shoulder and elbow joints.

Coordinating multiple segments at multiple joints to create linear motion of one part of the body is called *general motion*. Most human movement is general motion, and most of it is more complex than just tracing a straight line with a finger. Because it's complex, it's useful to look separately at the linear and angular motions that make up general motion.

Linear kinematics

Linear kinematics describes *linear motion,* or motion along a line (also called *translation*). There are two forms of linear motion:

- **Rectilinear motion:** Translation in a straight line. Your fingertip exhibited rectilinear motion as you successfully traced a line across the page or screen.

- **Curvilinear motion:** Translation along a curved line. Your fingertip exhibited curvilinear motion when you tried to move it across the page using only a single joint.

  Curvilinear motion also describes the path followed by an object moving through the air without support, like a thrown ball or a jumping child. This airborne body, whether it's a ball or a child, is called a *projectile*, and the curvilinear path it follows is called a *parabola* (an inverted U-shaped path).

Common descriptors of linear motion include how far the body moves, how fast the body moves, and the periods of slowing down or speeding up as it moves. Some familiar terms are used to describe linear motion, but in mechanics they have precise definitions:

- ✔ *Distance* and *displacement* are often used interchangeably to describe how far a body moves, but in mechanics *distance* simply means how far and *displacement* means how far in a specified direction.

- ✔ *Speed* and *velocity* both describe how fast a body moves, but in mechanics *speed* is simply how fast a body moves, while *velocity* refers to how fast the body moves in a specific direction.

- ✔ *Acceleration* is a tricky, but important, idea describing a change in velocity of a body. In everyday language, *acceleration* is often used to mean "speeding up" and *deceleration* is often used to mean "slowing down." In mechanics, *acceleration* is used to describe both speeding up *and* slowing down. The term is used both ways because acceleration provides a link between the description of motion, kinematics, and the force causing the motion, *kinetics*. For example, the force of gravity creates a downward acceleration on a body; when you jump into the air, the downward acceleration of gravity slows down your upward motion when you're going up, but speeds up your downward motion when you're coming down.

For more on all things related to linear kinematics, including projectiles and parabolic motion, jump right over to Chapter 5.

Angular kinematics

Angular kinematics describes *angular motion,* or motion involving rotations like swings, spins, and twists. Angular kinematics are used to describe the rotation of the whole body, like when a diver or gymnast performs a spin in the air, or the rotation of individual body segments, like when you bend or straighten your forearm at the elbow.

The common descriptors of angular motion include how far the body rotates, how fast the body rotates, and the periods of slowing down or speeding up while it rotates. The terms used to describe angular motion are similar to those used for linear kinematics, but they refer, as you might expect, to measures of angles.

- ✔ *Angular distance* and *angular displacement* describe how far a body rotates. Similar to linear kinematics, *angular distance* means how far the body rotates, while *angular displacement* means how far it rotates in a specified direction.

✔ *Angular speed* and *angular velocity* describe how fast a body rotates. *Angular speed* is just how fast the body rotates, but *angular velocity* refers to how fast it rotates in a specific direction.

✔ *Angular acceleration* is used to describe a change in the angular velocity of a body and can be used to describe both "speeding up" and "slowing down" the rate of rotation.

For more on all things related to angular kinematics, spin right over to Chapter 9.

Causing motion with kinetics

Kinetics is the subdivision of mechanics focused on the forces that act on a body to cause motion. Basically, a *force* is a push or a pull exerted by one body on another body. But a force, whether it's a push or a pull, can't be seen — we can see only the *effect* of a force on a body. An applied force wants to change the motion of the body — to speed it up or slow it down in the direction the force is applied. As I describe earlier, the speeding up or slowing down of a body is called *acceleration*.

Sir Isaac Newton formulated a set of three laws, appropriately called Newton's laws, describing the cause–effect relationship between the force applied and the changing motion, or acceleration, of a body. These three laws are the foundation for using kinetics to analyze both linear and angular motion. For more on Newton's laws, turn to Chapter 6.

Linear kinetics

Linear kinetics investigates how forces affect the linear motion, or *translation,* of a body. The characteristics of a force include its size, direction, point of application, and line of action. Each characteristic influences the force's effect on the body, and identifying the characteristics of each force applied to a body is an important step in kinetics. In Chapter 4, I show you how to describe the characteristics of a force and explain what makes gravity pull and friction push.

A body, especially the human body during movement, is usually acted on by several different external forces. The acceleration of the body is determined by the net force created by all the different forces acting at the same time. In Chapter 6, I show you the process of determining if the net force created by multiple forces represents an unbalanced, or unopposed, force; then I explain what Newton had to say about unbalanced force and why what he said is still important more than 300 years later.

From this basic understanding of unbalanced force and its effect on a body, you can use the impulse–momentum relationship to determine how an unbalanced force applied for a period of time speeds up or slows down the body.

Angular kinetics

Angular kinetics investigates the causes of angular motion, or *rotation*. The turning effect of a force applied to a body is called *torque*. Torque is produced when a force is applied to a body at some distance from an axis of rotation. I introduce the basic concept of torque in Chapter 8 and explain how the turning effect of a force is affected by manipulating the size of the force or by applying the force farther from the axis.

From this basic understanding of torque, I explain how muscle acts as a torque generator on the linked segments of the human body. The torque created by muscle interacts with the torque created by other external loads to cause, control, and stop the movement of segments.

Newton's laws make it possible to explain and predict the motion of all things. Using a Newtonian approach to analyze movement means to utilize the cause–effect relationship between the forces that act on a body and the motion of the body. Always.

Putting Biomechanics to Work

When you have the basic tools of kinematics and kinetics, along with a basic understanding of how the neuromusculoskeletal system controls movement, you can use them to analyze movement. In Part V, I show some common applications of using biomechanics to conduct an analysis:

- ✔ **Qualitative analysis:** This type of analysis is most frequently done in teaching, coaching, or clinical situations. You can apply the principles of biomechanics to visually evaluate the quality of a performance and provide feedback based on an accurate and specific troubleshooting of the cause of the level of performance.

- ✔ **Quantitative analysis:** This type of analysis measures kinematic and kinetic parameters of performance, usually using sophisticated laboratory equipment. It provides a more detailed description of a performance and is most typically used in a research study (or often in a laboratory experience in a biomechanics class).

- ✔ **Forensic analysis:** Biomechanics is one of the tools used to resolve criminal and civil legal questions. The principles of biomechanics are combined with evidence gathered by other investigators to answer the question of "whodunit."

Chapter 2

Reviewing the Math You Need for Biomechanics

. . . .

In This Chapter

▶ Setting a coordinate system

▶ Operating with algebra

▶ Dealing with the right triangle

▶ Working with vectors

. . . .

*M*athematics plays a big part in biomechanics. The equations in kine-matics, kinetics, and the mechanics of materials show the relation-ships among different factors that describe and explain motion and how materials react when loaded. From calculating the time for a fastball to reach home plate in baseball, through calculating the friction force provided by a new shoe design on a basketball court, to calculating the loading of bone during strength-training exercise, using math provides a tool for working in biomechanics.

Fortunately, a little dexterity with basic math skills goes a long way for suc-cess in biomechanics. The basic math skills include algebra, geometry, and a little bit of trigonometry. These provide the tools for solving problems in biomechanics, including working with the vector quantities in kinematics and kinetics.

In this chapter, I provide an overview of the basic math skills, with an empha-sis on using the skills in working with vectors.

Getting Orientated

To describe movement, you need a spatial reference system. The basic spatial reference system is the *Cartesian coordinate system,* shown in Figure 2-1. This two-dimensional coordinate system consists of two intersecting number lines; one creates a horizontal axis (often called the *x*-axis), and the other creates a vertical axis (often called the *y*-axis). Both the horizontal and vertical axes have positive and negative values. The arrows at the ends of the axes indicate that the axes extend forever.

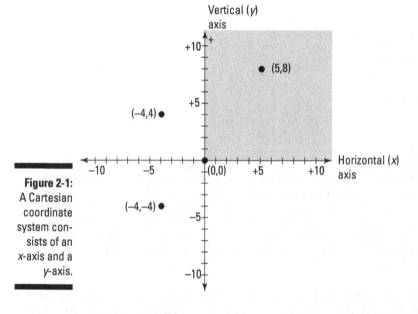

Figure 2-1: A Cartesian coordinate system consists of an *x*-axis and a *y*-axis.

Any location on the coordinate system is identified with a pair of coordinates using the format (*x,y*). The first value, *x*, describes how far the point is along the *x*-axis; the second value, *y*, describes how far the point is along the *y*-axis.

All spatial reference systems use a fixed *origin,* or starting point, from which the *x* and *y* measures are made. The origin is at the intersection of the horizontal and vertical axes and is designated (0,0) — it's 0 units along the *x*-axis, and 0 units along the *y*-axis.

Any point on the coordinate system can be identified with (*x,y*) coordinates. Figure 2-1 shows the point (5,8), which is 5 units along the *x*-axis and 8 units along the *y*-axis.

Although the *x*-axis and *y*-axis *look* like number lines, negative values are not less than positive values. Instead, the sign of a value on a reference system indicates *direction*. For the *x*-axis, positive values are to the right of the *y*-axis, and negative values are to the left of the *y*-axis. For the *y*-axis, positive values are above the *x*-axis, and negative values are below the *x*-axis.

Figure 2-1 shows the points (–4,–4) and (–4,4). The first point, (–4,–4) is 4 units to the left of the *y*-axis and 4 units below the *x*-axis. The second point, (–4, 4), is 4 units to the left of the *y*-axis and 4 units above the *x*-axis. Both points are 4 units away from the horizontal axis, but one point, (–4,–4) is 4 units below the *x*-axis and the other point, (–4,4), is 4 units above the *x*-axis.

To avoid using negative values in specifying a point's location, you can choose to set the coordinate system so that all points fall only in the positive quadrant. I do this as often as possible in this book.

Brushing Up on Algebra

Algebra is arithmetic without numbers. In arithmetic, the question may ask you to solve 5 + 3, dealing with two numbers. But in algebra, the question may ask you to solve $y = 3x + 3$, where *y* and *x* are variables and can take on different values. Many of the equations in biomechanics include variables, so this section provides an overview of the basic skills and terminology important for using algebra to solve equations.

Following the order of operations

The four basic operations in arithmetic are addition (+), subtraction (–), multiplication ($\cdot$ or $\times$), and division (/ or $\div$). Many problems with equation solving come from not knowing the order in which the operations are to be performed.

Consider this equation for the variable *x*, containing all the operations you'll be confronted with in biomechanics:

$$x = 5\sqrt{8 + \left(12 - 2^2\right)} + 4 \cdot 3 \div 2 + \frac{28}{7} + 3(7 + 3 \cdot 10 \div 5 + 1)$$

Three rules specify the order in which the mathematical operations *must* be performed:

✔ Consider all terms within parentheses as one term, and solve these first.

✔ Do all exponents (powers or roots) next.

✔ Always multiply and divide *before* adding and subtracting, moving from left to right.

The order of operations can be remembered by the acronym PEMDAS, where *P* stands for parentheses, *E* stands for exponents (powers or radicals), *M* stands for multiplication, *D* stands for division, *A* stands for addition, and *S* stands for subtraction. So, it looks like this:

1. Solve all computations within **Parentheses**.

2. Solve all **Exponents** (powers or radicals).

3. Solve all **Multiplication**.

4. Solve all **Division**.

5. Solve all **Addition**.

6. Solve all **Subtraction**.

By following PEMDAS, you can solve even the most complex-looking equation. For example, apply PEMDAS to the equation for *x* listed earlier:

$$x = 5\sqrt{8 + \left(12 - 2^2\right)} + 4 \cdot 3 \div 2 + \frac{28}{7} + 3(7 + 3 \cdot 10 \div 5 + 1)$$

1. Solve all computations within parentheses.

Moving from left to right, the first set of parentheses you see is $(12 - 2^2)$, but wait! That set of parentheses has a number with an exponent inside and it's inside a radical, so let's skip that one and continue moving to the right.

The next set of parentheses you see is: $(7 + 3 \cdot 10 \div 5 + 1)$. Within this parentheses, there are several operations to perform. PEMDAS says that you should always do multiplication and division before addition and subtraction. So, moving from left to right, the first sign is a plus sign (+), so you can skip that. The next sign is a multiplication sign (·), so work that out: $3 \cdot 10 = 30$. Now, you have $(7 + 30 \div 5 + 1)$.

Continuing on toward the right, the next sign is a division sign (÷), so work that out: $30 \div 5 = 6$. Now, you have $(7 + 6 + 1)$. You're left with all addition symbols, so finish it off, and you get 14 inside the parentheses.

The equation now looks like this:

$$x = 5\sqrt{8 + \left(12 - 2^2\right)} + 4 \cdot 3 \div 2 + \frac{28}{7} + 3(14)$$

2. Solve all exponents (powers or radicals).

Moving from left to right, the first exponent you see is the radical $\sqrt{8 + \left(12 - 2^2\right)}$. Under the radical sign, there are several operations to perform. PEMDAS says that you should start with parentheses, where you have $(12 - 2^2)$. You need to solve all exponents before you can do anything else, so $2^2 = 4$. That leaves you with $\sqrt{8 + (12 - 4)} = \sqrt{8 + 8} = \sqrt{16}$. The square root of 16 is 4.

The equation now looks like this:

$$x = 5(4) + 4 \cdot 3 \div 2 + \frac{28}{7} + 3(14)$$

3. **Solve all multiplication.**

Moving from left to right, the first multiplication you have is 5(4), which is 20. The second multiplication you see is $4 \cdot 3$, which is 12. And the third multiplication you see is 3(14), which is 42.

The equation now looks like this:

$$x = 20 + 12 \div 2 + \frac{28}{7} + 42$$

4. **Solve all division.**

Moving from left to right, the first division you have is $12 \div 2$, which is 6. And the next division you see is $\frac{28}{7}$, which is 4.

The equation now looks like this:

$$x = 20 + 6 + 4 + 42$$

5. **Solve all addition.**

You're in the home stretch now. Add up the numbers, and you get $x = 72$. (You got away without having to do any subtraction at the end!)

Defining some math operations

When you're working on math problems in biomechanics, you'll encounter words like *sum, difference, product,* and *quotient,* and you need to know what those words mean in order to do the math correctly. In this section, I walk you through these words and give you some examples.

Sum

The *sum* is the quantity that results from adding other quantities. For example, $10 + 2 = 12$. Or, to put it differently, 12 is the *sum* of 10 and 2. Here's another example: $a = b + c$. Or, to put it differently, a is the *sum* of b and c.

Difference

The *difference* is the quantity that results from subtracting other quantities. For example, $10 - 2 = 8$. Or, to put it differently, 8 is the *difference* between 10 and 2. Here's another example: $a = b - c$. Or, to put it differently, a is the *difference* between b and c.

Product

The *product* is the quantity that results from multiplying other quantities. For example, $10 \cdot 2 = 20$. Or, to put it differently, 20 is the *product* of 10 and 2. Here's another example: $a = b \cdot c$. Or, to put it differently, a is the *product* of b and c.

Quotient

The *quotient* is the quantity that results from dividing other quantities. For example, $10 \div 2 = 5$. Or, to put it differently, 5 is the *quotient* of 10 and 2. Here's another example: $a = b \div c$. Or, to put it differently, a is the *quotient* of b and c.

Isolating a variable

Given the equation $x = 6 + 7$, solving for x is pretty straightforward: Simply add the terms on the other side of the equal sign (=) from x, and you get 13. Unfortunately, not all equations are that easy. Often, the unknown variable is buried on one side of the equation among several other terms, the way x is in the following equation:

$$8 + 5x - 2 = 9 + 7 \cdot 14$$

In such a case, you have to isolate x on one side of the equal sign, all by itself. You do this by *transposing* terms (moving terms from one side of the equal sign to the other).

To transpose a term, you subject *both* sides of the equation to the term's *inverse* operation. For instance, in the equation above, to transpose the 8, you *subtract* 8 from both sides of the equation. Similarly, to transpose –2, you *add* 2 to both sides. Finally, to reduce $5x$ to x, you *divide* both sides by 5. After completing these operations, x will be isolated on one side of the equation.

In mathematical form, these steps look like this:

$$8 + 5x - 2 = 9 + 7 \cdot 14$$
$$8 - 8 + 5x - 2 = 9 + 7 \cdot 14 - 8$$
$$5x - 2 = 9 + 7 \cdot 14 - 8$$
$$5x - 2 + 2 = 9 + 7 \cdot 14 - 8 + 2$$
$$5x = 9 + 7 \cdot 14 - 8 + 2$$
$$\frac{5x}{5} = \frac{9 + 7 \cdot 14 - 8 + 2}{5}$$
$$x = \frac{9 + 7 \cdot 14 - 8 + 2}{5}$$
$$= \frac{9 + 98 - 8 + 2}{5}$$
$$= \frac{101}{5}$$
$$= 20.2$$

Whenever you transpose a term, the operation you perform on one side of the equal sign gets performed on the other side of the equal sign.

Sometimes in biomechanics, you'll be able to use a specific equation to solve a given problem. For example, the equation $v^2 = u^2 + 2ap$ defines the final velocity of a body (v) in terms of its initial velocity (u), acceleration (a), and displacement (p). (I explain this equation in Chapter 5, along with each of the terms I just mentioned.) If values for three of the four terms (called the *known values*) are given in the problem, you can solve for the fourth term (called the *unknown value*).

Unless the problem asks to solve for v, which is already isolated on the left side of the equation, you need to isolate the unknown value and then plug in the known values to solve the equation. For example, if the unknown value is p, the displacement, use the process I just described to isolate for p in the equation. First, to transpose the u^2, subtract u^2 from both sides of the equation, and you get

$$v^2 - u^2 = 2ap$$

Next, to transpose $2a$, divide both sides by $2a$, and you get

$$\frac{v^2 - u^2}{2a} = p$$

If you want, you can rearrange it to the following:

$$p = \frac{v^2 - u^2}{2a}$$

With p isolated, you can substitute the values for v, u, and a into the equation, apply PEMDAS, and calculate the value of p.

Interpreting proportionality

In math, variables are proportional when a change in one variable is always matched by a change in the other variable. There are two rules to interpreting proportionality:

- ✔ **When an increase in one variable increases the value of the other, the variables are *directly proportional*.** A directly proportional relationship is written in the general form of $y = x$. To maintain the equality of the equation, as x gets bigger, y gets bigger. So, if $x = 2$, then $y = 2$. If $x = 4$, then $y = 4$.

- ✔ **When a decrease in one variable decreases the value of the other, the variables are *inversely proportional*.** An inversely proportional relationship is written in the general form of $y = \frac{1}{x}$. To maintain the equality of the equation, as x gets bigger, y gets smaller. So, if $x = 2$, then $y = \frac{1}{2}$. If $x = 4$, then $y = \frac{1}{4}$.

Consider the equation $F = ma$, a simplistic statement of an important law in mechanics known as Newton's second law of motion (you can find more detail on this and Newton's other two laws of motion in Chapter 6). Isolating for a, the equation becomes

$$a = \frac{F}{m}$$

The terms on the right side of the equation, F and m, are proportional to a. Both F and m affect the size of a.

Using the rules above, a is *directly* proportional to F. If m is a constant value (say, 10), as F gets larger, so does a. So, if $F = 20$, then $a = 2$. If $F = 30$, then $a = 3$.

Conversely, a is *inversely* proportional to m. If F is a constant value (say, 100), as m gets larger, a gets smaller. So, if $m = 20$, then $a = 5$. If $m = 100$, then $m = 1$.

Looking for the Hypotenuse

Some basic principles from geometry are very important in biomechanics. The right-angle triangle (often just called a right triangle) is a useful figure when working problems in biomechanics, because the relationships among the lengths of the sides of the right triangle provide a toolbox for working with vectors.

Figure 2-2 shows a right triangle. In a right triangle, one angle is a right angle, measuring 90 degrees. The 90-degree angle is indicated with a small square at the angle.

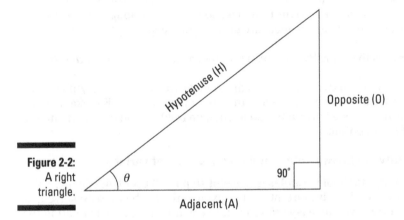

Figure 2-2:
A right
triangle.

The side across from the 90-degree angle in a right triangle is called the *hypotenuse.* The hypotenuse is always the longest side of the triangle.

In all triangles, the sum of the three angles is 180 degrees. Because one angle in a right triangle is 90 degrees, the sum of the other two angles is obviously 90 degrees. Each of these two other angles is less than 90 degrees; angles less than 90 degrees are called *acute angles.*

In Figure 2-2, one of the acute angles is set as the *reference angle;* it's indicated with θ (the Greek letter theta). The reference angle is used to give names to the other two sides of the right triangle. The side across from θ is called the opposite (O) side. The side meeting with the hypotenuse to create θ is called the adjacent (A) side — it's adjacent to the reference angle (θ).

The basic step of choosing one of the acute angles to be the reference angle (θ) and naming the two sides as opposite and adjacent is a critical step in using the right triangle to solve problems in biomechanics. The hypotenuse will never change (it's always the longest side), but the side you call opposite or adjacent will change depending on which θ you're interested in.

Using the Pythagorean theorem

The Pythagorean theorem relates the length of the three sides of a right triangle. Using the names of the sides of the right triangle in Figure 2-2, the equation for the Pythagorean theorem is written as $\text{Opposite}^2 + \text{Adjacent}^2 = \text{Hypotenuse}^2$ (which can be abbreviated to $O^2 + A^2 = H^2$). The Pythagorean theorem holds true for all right triangles, no matter how long or short the sides are, and no matter the measure of the acute angles.

$O^2 + A^2 = H^2$ is the same Pythagorean theorem as the more familiar $a^2 + b^2 = c^2$.

When the lengths of two sides of a right triangle are known, the Pythagorean theorem is used to calculate the length of the unknown side. For example, if the opposite side is 30 m and the adjacent side is 40 m, you can calculate the length of the hypotenuse.

1. **Identify the known values and create a table of variables.**

   Creating a table of variables is a useful step in all problem solving. You simply list all the important values given in the problem to be solved, which gives you an easy reference for selecting the correct equation to apply.

   In this example, here's the table of variables you would come up with:

   - Hypotenuse (H) = Unknown
   - Opposite side (O) = 30 m
   - Adjacent side (A) = 40 m

2. **Select the equation based on the unknown variable you're trying to solve for.**

   In this case, the correct equation is the Pythagorean theorem, $O^2 + A^2 = H^2$.

3. **Isolate the unknown value, and substitute the known values.**

   The unknown value is H, so isolate that first by taking the square root of both sides:

$$H^2 = O^2 + A^2$$
$$H = \sqrt{O^2 + A^2}$$
$$= \sqrt{30^2 + 40^2}$$
$$= \sqrt{900 + 1,600}$$
$$= \sqrt{2,500}$$
$$= 50 \text{ m}$$

   The length of the hypotenuse is 50 m.

It's good practice to show the steps you use to solve the equation, as I did above. If your answer is wrong, by showing the steps you used, you can more easily go back and see where you made a mistake. I use this step-by-step process of solving problems throughout this book, and I recommend you use a similar step-by-step process for all problem solving — whether it's the one I use or one that you adopt on your own or learn from your teacher.

Let's try another problem. If given the length of the hypotenuse (100 m) and the length of the adjacent side (25 m), calculate the length of the opposite side. Use the same steps as you did before:

1. **Identify the known values and create a table of variables.**

   - Hypotenuse (H) = 100 m
   - Opposite side (O) = Unknown
   - Adjacent side (A) = 25 m

2. **Select the equation based on the unknown variable you're trying to solve for.**

   In this case, the correct equation is the Pythagorean theorem, $O^2 + A^2 = H^2$.

3. **Isolate the unknown value, and substitute the known values.**

   The unknown value is O, so:

$$O^2 + A^2 = H^2$$
$$O^2 = H^2 - A^2$$

Now, take the square root of both sides and substitute the known values:

$$O = \sqrt{H^2 - A^2}$$
$$= \sqrt{100^2 - 25^2}$$
$$= \sqrt{10,000 - 625}$$
$$= \sqrt{9,375}$$
$$= 96.8 \text{ m}$$

The length of the opposite side is 96.8 m.

De-tricking trigonometric functions: SOH CAH TOA

Trigonometry is the field of mathematics describing the relationship between the sides and angles of triangles. Just as the Pythagorean theorem (see the preceding section) allows you to calculate the length of an unknown side of a right triangle when the lengths of two sides are known, the trigonometic functions (or simply trig functions) allow you to calculate the length of the sides of a right triangle if the length of one side and the measure of one acute angle are known. The trig functions expand your toolbox for working with right triangles.

Defining the trig functions

The trig functions specify the ratios between two sides of the right triangle. A *ratio* is the relationship between two measurable quantities, expressed in the format

$$\text{ratio} = \frac{\text{one measurement}}{\text{another measurement}}$$

Or, for a right triangle:

$$\text{trig function} = \frac{\text{length of one side}}{\text{length of another side}}$$

The ratios are called trig functions because the relationship between the lengths of the sides of a right triangle depends on, or is a function of, the size of the specified acute angle.

The three trig functions of interest are the sine, the cosine, and the tangent. Using the sides of the labeled right triangle in Figure 2-2, the three trig functions are defined as follows:

$$\sin\theta = \frac{\text{Opposite}}{\text{Hypotenuse}} = \text{SOH}$$

$$\cos\theta = \frac{\text{Adjacent}}{\text{Hypotenuse}} = \text{CAH}$$

$$\tan\theta = \frac{\text{Opposite}}{\text{Adjacent}} = \text{TOA}$$

Each trig function defines the ratio between two sides of a right triangle as a function of the angle θ. Using the sine as an example, each function can be read as "The sine function specifies the ratio between the opposite side and the hypotenuse as a function of the measure of the acute angle θ." The ratio between the length of the opposite side and the length of the hypotenuse is always the same for a given angle θ, regardless of the actual length of the two sides. The same holds for the other trig functions.

You can remember the three trig functions using the letters SOH CAH TOA, which is short for the first letter of the trig function and the first letter of the two sides defined by the function.

Using the trig functions to calculate lengths

A right triangle with a hypotenuse of 15 m and one acute angle of 74 degrees is shown in Figure 2-3. Setting the 74-degree angle as θ, the adjacent and opposite sides are labeled. To solve the length of these sides, start by calculating the length of the adjacent side:

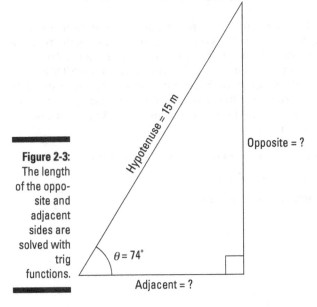

Figure 2-3: The length of the opposite and adjacent sides are solved with trig functions.

Hypotenuse = 15 m

Opposite = ?

$\theta = 74°$

Adjacent = ?

1. **Identify the known values and create a table of variables.**

   - $\theta = 74$ degrees
   - Hypotenuse (H) = 15 m
   - Adjacent side (A) = Unknown

2. **Select the trig function to use.**

   A process of elimination makes choosing the correct function easier. First, list out SOH CAH TOA. Then compare each function to the table of variables to choose the correct one.

   SOH needs the angle θ, which you know, the hypotenuse, which you also know, and the opposite side, which you don't know, so SOH can't be used.

   CAH needs the angle θ, which you know, the hypotenuse, which you know, and the adjacent side, which you don't know. So, cosine is the trig function to use.

3. **Expand the chosen function, isolate for the unknown value, and solve.**

   You know that $\cos\theta = \dfrac{\text{Adjacent}}{\text{Hypotenuse}}$, and the unknown in this problem is the adjacent side, so isolate that:

   $$\text{Adjacent} = \cos\theta \cdot \text{Hypotenuse} = \cos 74° \cdot 15 \text{ m}$$

   Whoa! What's the value of cos 74°? Pull out your trusty calculator, enter 74, and press the COS key. (On some calculators, you have to press COS and then enter 74.) The value of 0.2756 (with some trailing values) should appear. You can round this to 0.28 for the calculations.

WARNING!

Make sure your calculator is in the correct angle mode when using the trig functions. The two angle modes on most calculators are degrees and radians. You'll see a MODE button of some sort on your calculator (check your manual if you're not sure). When you press the MODE button, it shows whether you're in Deg mode (for degrees) or Rad mode (for radians). Make sure you're in the Deg mode when the angle is measured in degrees, and Rad when the angle is measured in radians, or your answers will be wrong.

Enter the value of the cosine of 74 degrees in the equation:

   $$\text{Adjacent} = 0.28 \cdot 15 \text{ m} = 4.2 \text{ m}$$

The length of the adjacent side is 4.2 m.

To calculate the length of the opposite side, follow the same steps you'll come to know and love:

1. **Identify the known values and create a table of variables.**

   - $\theta = 74$ degrees
   - Hypotenuse (H) = 15 m
   - Opposite side (O) = Unknown

2. **Select the trig function to use.**

   First, list out SOH CAH TOA, and compare each function to the table of variables to choose the correct function.

   SOH needs the angle θ, which you know, the hypotenuse, which you know, and the opposite side, which you don't know. So, sine is the trig function to use.

3. **Expand the chosen function, isolate for the unknown value, and solve.**

   You known that $\sin\theta = \dfrac{\text{Opposite}}{\text{Hypotenuse}}$, and the unknown in this problem is the opposite side, so isolate that:

   $$\text{Opposite} = \sin\theta \cdot \text{Hypotenuse} = \sin 74° \cdot 15\text{ m} = 0.96 \cdot 15\text{ m} = 14.4\text{ m}$$

   The length of the opposite side is 14.4 m.

You can check that the lengths are correct by plugging the values into the Pythagorean theorem, $A^2 + O^2 = H^2$:

$$4.2^2 + 14.4^2 = 15^2$$
$$17.6 + 207.4 = 225$$
$$225 = 225$$

Correctly labeling the sides of the right triangle is important. If they're reversed, your calculated lengths will be right but they'll be assigned to the wrong sides of the triangle.

In biomechanics, the tangent function is directly used less often than the sine and cosine functions. I won't show an example using the tangent function, but the steps used are the same as those for sine and cosine. A version of the tangent function is frequently used, however, and I show this application in the next section.

Using the arctangent to solve for the angle θ

In some situations, you know the values of the adjacent side and the opposite side, but you don't know the value of the hypotenuse or of angle θ. This situation is shown in Figure 2-4, with an adjacent side of 270 m and an opposite side of 180 m.

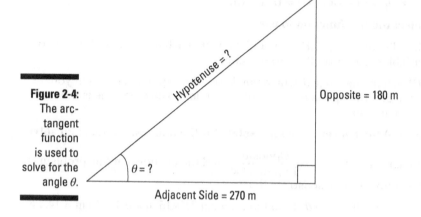

Figure 2-4: The arctangent function is used to solve for the angle θ.

In this example, you can calculate the length of the hypotenuse using the Pythagorean theorem:

$$H^2 = A^2 + O^2$$
$$H = \sqrt{A^2 + O^2}$$
$$= \sqrt{270^2 + 180^2}$$
$$= \sqrt{105,300}$$
$$= 324.5 \text{ m}$$

When you know the adjacent and opposite sides, but not θ, you can use a trigonometric function known as the arctangent, or arctan, to solve for θ. The arctan function is also known as the *inverse tangent*. It's called "inverse" because the length of the two sides are known, giving the ratio between them, and the ratio is used to find the angle θ. (In the use of sine and cosine earlier, the angle was known and the goal was to calculate the length of the opposite or adjacent side from the relationship with the hypotenuse.)

The equation for the arctan is:

$$\theta = \arctan\left(\frac{\text{Opposite}}{\text{Adjacent}}\right)$$

Think of this equation as "What angle has the tangent function equal to the opposite side divided by the adjacent side?" (Note that the ratio, $\frac{\text{Opposite}}{\text{Adjacent}}$, is the same as for the tangent.)

To calculate the angle θ, follow these steps:

1. **Identify the known values and create a table of variables.**

   - θ = Unknown

   - Adjacent side (A) = 270 m

   - Opposite side (O) = 180 m

2. **Select the equation for arctangent, and substitute the given values.**

$$\theta = \arctan\left(\frac{\text{Opposite}}{\text{Adjacent}}\right)$$
$$= \arctan\left(\frac{180}{270}\right)$$
$$= \arctan(0.67)$$

To use your calculator, enter 0.67 and press the ATAN key, or press ATAN and then enter 0.67. (On some calculators, the key is identified as $\tan^{-1}$; on others, you press the INV key and then TAN. Consult your user's manual.)

The value of 33.8 (with some trailing values) should appear. If the value 0.59 appears, you're in Radian mode. Switch the mode to Degrees and renter the values.

The angle θ is 33.8 degrees. This is the angle between the hypotenuse and the adjacent side of the right triangle.

Unvexing Vector Quantities

A *quantity* is anything that can be measured. In biomechanics, two types of quantities are important:

- ✔ **Scalars:** A *scalar quantity* is any quantity that can be fully described by its magnitude, size, or amount. Scalar quantities include time, mass, distance, and speed. A scalar quantity is fully described by its magnitude, the specific number of units used to measure it — for example, time = 8 seconds (or s), mass = 75 kg, distance = 2 m, or speed = 20 m/s.

- ✔ **Vectors:** A *vector quantity* is described not just by its magnitude but also by a direction associated with the quantity. Vector quantities include force, displacement, velocity, and acceleration. Vector quantities are fully described only if *both* magnitude and direction are specified — for example, force = 20 Newtons (or N) to the right, displacement = 5 m at an angle of 40 degrees to the horizontal, velocity = 5 m/s at an angle of 20 degrees to the horizontal, or acceleration = 10 m/s/s downward.

A vector quantity is graphically represented by an arrow called, simply, a *vector.* A vector is drawn so the magnitude of the quantity is represented by the length of the arrow, and the direction of the quantity is represented by the tip, or *arrowhead.*

Vectors are drawn on an arbitrary coordinate system of an *x*-axis and a *y*-axis (or *xy*-coordinate system). Figure 2-5 shows the graphic representation of the displacement vector of 5 m at an angle of 40 degrees drawn on an *xy*-coordinate system. The 40-degree angle of its direction is measured from the right horizontal (unless otherwise specified). Sometimes, only part of the coordinate system is drawn, as in the middle part of Figure 2-5. The complete coordinate system is inferred to be present.

Figure 2-5:
Vector representation
of a displacement
of 5 m at an
angle of 40
degrees.

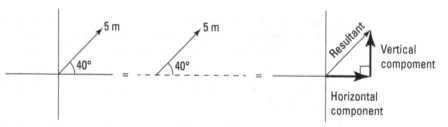

The coordinate system can be drawn at any orientation that makes sense for the situation described. For example, when dealing with the human body, the coordinate system is often set at a joint of interest, with the *y*-axis aligned to the long axis of the bone and the *x*-axis aligned across the bone. In every case, the directions of *x* and *y* must be specified when setting the coordinate system.

Drawing vectors on a coordinate system shows that vectors can be considered to consist of two components, or parts, acting at a right angle (90 degrees) to each other. The components of the displacement of 5 m are shown in the right part of Figure 2-5. The original vector (5 m at 40 degrees) is called the *resultant vector,* because it results from the combination of the two components. The two components are called the *horizontal component* and the *vertical component.*

Resolving a vector into components

An important tool in biomechanics is being able to calculate the magnitude of the two components of a vector, called *resolving a vector* or *vector resolution*.

A resultant vector and its two components acting at right angles create a right triangle. Because a right triangle is created, the trigonometric functions explained earlier in this chapter are used for resolving a vector into components.

Figure 2-6 shows the resultant displacement vector of 5 m at an angle of 40 degrees and the two components as a right triangle. It's simply the right part of Figure 5-5 redrawn with the sides labeled using the 40-degree angle specified as θ. The resultant vector is the hypotenuse, the horizontal component is the adjacent side, and the vertical component is the opposite side.

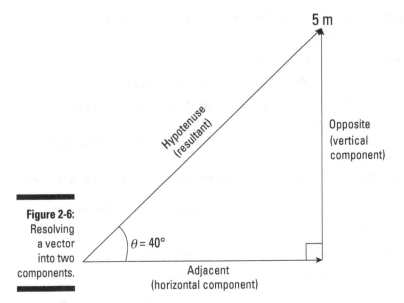

Figure 2-6: Resolving a vector into two components.

To calculate the length of the horizontal component (the adjacent side), follow these steps:

1. **Identify the known values and create a table of variables.**

   - θ = 40 degrees
   - Hypotenuse (H) = resultant vector = 5 m
   - Adjacent side (A) = horizontal component = Unknown

2. Use the process of elimination to select the trig function to use.

You can eliminate SOH because you don't know the opposite side. CAH needs the angle θ, which you know, the hypotenuse, also known, and the adjacent side, which you're trying to find. Select cosine as the trig function to use.

3. Expand the chosen function, isolate for the unknown value, and solve.

You know that $\cos\theta = \dfrac{\text{Adjacent}}{\text{Hypotenuse}}$, and the unknown in this problem is the adjacent side, so isolate that:

$$\text{Adjacent} = \cos\theta \cdot \text{Hypotenuse} = \cos 40° \cdot 5\text{ m} = 0.77 \cdot 5\text{ m} = 3.8\text{ m}$$

The adjacent side, or horizontal component, is 3.8 m.

Now calculate the length of the vertical component (the opposite side):

1. Identify the known values and create a table of variables.

- $\theta = 40$ degrees

- Hypotenuse (H) = resultant vector = 5 m

- Opposite side (O) = vertical component = Unknown

2. Use the process of elimination to select the trig function to use.

SOH needs the angle θ, which you know, the hypotenuse, which you know, and the opposite side, which you're trying to find. Select sine as the trig function to use.

3. Expand the chosen function, isolate for the unknown value, and solve.

You know that $\sin\theta = \dfrac{\text{Opposite}}{\text{Hypotenuse}}$, and the unknown in this problem is the opposite side, so isolate that:

$$\text{Opposite} = \sin\theta \cdot \text{Hypotenuse} = \sin 40° \cdot 5\text{ m} = 0.64 \cdot 5\text{ m} = 3.2\text{ m}$$

The opposite side, or vertical component, is 3.2 m.

So, the horizontal component is 3.8 m and the vertical component is 3.2 m.

The same steps to resolve a vector into components can be applied to any vector. The key steps are to visualize the resultant vector as the hypotenuse of a right triangle and to correctly label the components as the adjacent and opposite sides relative to the reference angle θ.

Composing a vector from components

Another important tool in biomechanics is to calculate the magnitude and direction of the resultant vector given the magnitude of the two components. This is called *composing a vector* or *vector composition.* In this section, I show you how to compose the resultant force vector when a horizontal force of 20 N and a vertical force of 100 N are applied to a body.

Because the two components always act at 90 degrees to each other, the resultant vector and the two components can be considered to form a right triangle (shown in Figure 2-7). In A, the two component force vectors are shown by themselves. In B, lines drawn parallel to and equal in length to the horizontal and vertical forces create a rectangle. In C, a diagonal line drawn across the rectangle creates two right triangles. In D, one of the right triangles is isolated and the sides are labeled. The diagonal of the rectangle is the hypotenuse, the resultant force from the two components. The angle of the resultant is labeled with θ, and the two components are labeled as opposite and adjacent relative to θ. Because a right triangle is created, the Pythagorean theorem and the trigonometric function arctan (covered earlier) are used to calculate the magnitude and direction of the resultant vector.

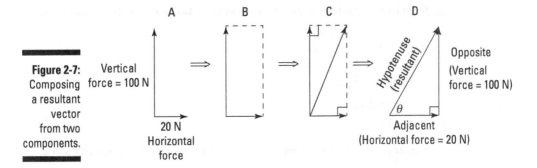

Figure 2-7: Composing a resultant vector from two components.

Calculate the magnitude of the resultant force (the hypotenuse) using the Pythagorean theorem:

1. **Identify the known values and create a table of variables.**

    • Hypotenuse (H) = resultant force = Unknown

    • Opposite side (O) = vertical component = 100 N

    • Adjacent side (A) = horizontal component = 20 N

2. **Select the equation, isolate the unknown value, and substitute the known values.**

$$H^2 = O^2 + A^2$$
$$H = \sqrt{O^2 + A^2}$$
$$= \sqrt{100^2 + 20^2}$$
$$= \sqrt{10,000 + 400}$$
$$= \sqrt{10,400}$$
$$= 102 \text{ N}$$

The magnitude of the resultant force (the hypotenuse) is 102 N.

Calculate the angle θ, the direction of the resultant force, using arctan:

1. **Identify the known values and create a table of variables.**

   - $\theta =$ Unknown
   - Adjacent side (A) = horizontal force = 20 N
   - Opposite side (O) = vertical force = 100 m

2. **Select the equation for arctan, substitute the given values, and solve.**

$$\theta = \arctan\left(\frac{\text{Opposite}}{\text{Adjacent}}\right)$$
$$= \arctan\left(\frac{100}{20}\right)$$
$$= \arctan(5)$$
$$= 78.7°$$

The angle θ is 78.7 degrees. This is the direction of the resultant force.

The resultant force is 102 N at an angle of 78.7 degrees.

Chapter 3

Speaking the Language of Biomechanics

*B*iomechanics involves applying the principles of mechanics to a living body. Sometimes we even use biomechanics to study nonliving things, such as baseball bats, golf balls, and running shoes. But what exactly is mechanics? *Mechanics* is the science studying the effects of forces on a body. *Biomechanics* is a tool for objectively observing, evaluating, and correcting movement to improve performance and reduce the risk of injury.

As in all sciences, terminology is important in biomechanics. Using the precise mechanical definition of a term standardizes the description of movement. Fortunately, most of the terms that make up the mechanics of biomechanics — including force, speed, velocity, acceleration, impulse, energy, and strain — are probably already in your vocabulary. Similarly, most of the terms in the "bio" part of biomechanics are familiar — including bone, muscle, and nerve. All these terms have a specific meaning in biomechanics, and this is important because it allows consistency and specificity when talking about the body under analysis.

In this chapter, I define a variety of terms used in biomechanics. Throughout this book, I expand on and clarify the definitions as the terms are used and applied to movement.

Measuring Scalars and Vectors

Two important terms in biomechanics are *scalar* and *vector*. These terms refer to the characteristics of something *measured*, or quantified by giving it a numeric value. And a lot of things are measured in biomechanics.

A *scalar quantity* is something described fully by providing a measure of "how much," or the magnitude of the quantity. One very important scalar quantity is *mass*, the measure of how much matter a body contains. In biomechanics, mass is typically measured in kilograms (kg). Other scalar quantities include distance and speed, both measures of the motion of a body.

A *vector quantity* is only fully described if both its magnitude and its direction are reported. One of the most important vector quantities in biomechanics is *force*, a push or a pull that tends to cause a change in the motion of a body. Other vector quantities include displacement, velocity, and acceleration, the basic measures in *kinematics* (the description of the motion of a body).

Vector quantities are represented using arrows, with the tip of the arrow showing the direction of the quantity and the length of the arrow representing the magnitude of the quantity. Two basic tools for working with vectors in biomechanics are

- ✔ **Vector resolution:** Vectors can be resolved into components, typically a horizontal component and a vertical component (often called the x and y components) acting at 90 degrees to each other. This process, called *vector resolution,* applies the trigonometric functions of the sine, cosine, and tangent. In Chapter 2, I explain the trigonometric functions and how to use the easy-to-remember anagram SOH CAH TOA when resolving a vector into two components.

- ✔ **Vector composition:** *Vector composition* allows for the combination of two separate vectors into a single resultant vector (resultant because it represents the combined effect of the two vectors). In Chapter 2, I explain this process and show how vector composition uses the Pythagorean theorem and the inverse tangent trigonometric function to compose a resultant from the vectors.

The process of vector composition can be extended to many vectors by first resolving each into its horizontal and vertical components, adding all the components in the horizontal direction and all the components in the vertical direction, and then applying the same steps as when working with two orthogonal vectors.

Standardizing a Reference Frame

The whole body and the individual segments of the human body can move in many directions, creating a need to standardize the description of movement. For example, if you're told to raise your arm, you can lift it straight out in front of your body, or lift it out sideways from your body. Both actions raise your arm, but it's confusing if one instruction can cause two different movements of the arm.

Figure 3-1 shows a human body in the *anatomical position,* standing upright, with the arms held out slightly from the sides with the palms of the hands facing forward, and with the feet about shoulder width apart. The anatomical position is the standard reference position for describing the body, although it's not a natural standing position.

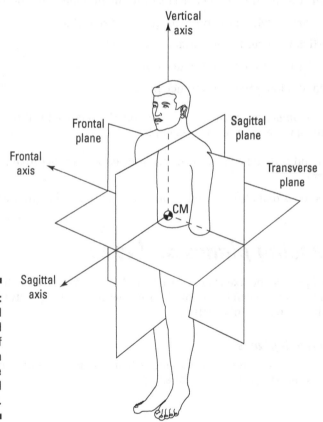

Figure 3-1: The cardinal planes and axes of the human body in the anatomical position.

Directing your attention to locations of the body

The following directional terms are typically used to specify the relative location of different body segments, muscles, landmarks on a bone, and other tissues:

- **Anterior:** In front of, or toward the front of the body
- **Posterior:** Behind, or toward the back of the body
- **Inferior:** Below or lower, or farther from the head
- **Superior:** Above or higher, or closer to the head
- **Lateral:** Toward the side, or farther from the midline of the body
- **Medial:** Toward the middle, or closer to the midline of the body
- **Deep:** Farther within the body from the outer surface
- **Superficial:** Closer to the surface of the body
- **Distal:** Farther from the reference point
- **Proximal:** Closer to the reference point

Distal and *proximal* are usually used to identify the relative location of joints or segments of the arm or leg. For example:

- Your hand is distal to your forearm. Your upper arm is proximal to your forearm.
- Your knee is distal to your hip. Your knee is proximal to your ankle.

Referencing planes and axes

A system of planes and axes is used to describe movement of the human body (and the system can be adapted for use with any living body). I describe this system in this section.

Moving through planes

A *plane* is a two-dimensional flat surface. When a body segment moves, it's described as "moving through a plane."

There are three standard planes for describing movement of the body, as shown in Figure 3-1:

- ✔ **Sagittal plane:** The sagittal plane divides the body into right and left parts. When you look at someone from the side, you see them in the sagittal plane.

- ✔ **Frontal plane:** The frontal plane divides the body into front (anterior) and back (posterior) parts. When you're standing in front of or behind someone, you see her in the frontal plane.

- ✔ **Transverse plane:** The transverse plane divides the body into an upper (superior) and a lower (inferior) part. Sometimes the transverse plane is called the *horizontal plane,* especially when the body is upright. When you're looking at someone from above, a bird's-eye view, you see her in the transverse plane.

The planes are aligned at right-angles to each other. When a standard plane passes through the center of gravity of the body it's called a *cardinal plane.* The center of gravity of the body, which I explain in Chapter 8, is called the *balance point* of the body. A *diagonal plane* is a plane cutting across the cardinal planes at an angle. There are an infinite number of ways an oblique plane can cut through a body.

Obviously, there are no physical planes cutting through the body, so we say the planes are "imaginary." Using the three planes conveniently provides a common and universal descriptor of any whole body or segment movement that occurs.

Rotating at an axis

An *axis* is a real or imaginary point around which rotation occurs. When something rotates, it's described as "rotating around an axis." An example of a real axis is the axle of a bicycle wheel. An example of an imaginary axis is a line passing through the elbow joint around which the forearm rotates as the elbow bends and straightens.

An axis of rotation is perpendicular, or at 90 degrees, to the plane through which the body or segment moves as it rotates. The same axis will always have the same plane paired with it. Here are the axes of the body (shown in Figure 3-1):

- ✔ **Vertical axis:** A vertical axis is aligned from the top to the bottom of the body. When a body rotates around a vertical axis, the movement of the body is through the transverse plane. When you look at someone from above, your eye is looking along a vertical axis. When a vertical axis is applied to an individual segment, it's set to run the length of the segment and is called a *longitudinal axis*.

✔ **Frontal axis:** A frontal axis goes from side to side across the body, aligned within the frontal plane. Another common name for the frontal axis is the *mediolateral axis*. When a body rotates around a frontal axis, it moves through the sagittal plane. When you look at someone from the side, your eye is looking along a frontal axis.

✔ **Sagittal axis:** A sagittal axis runs along the sagittal plane, going from the front to the back (or the back to the front) of the body. A sagittal axis is sometimes called an *anteroposterior axis*. When a body rotates around a sagittal axis, it moves through the frontal plane. When you look at someone from in front or behind, your eye is looking along a sagittal axis.

Describing Movement: Kinematics

Kinematics is the branch of mechanics that describes *motion* (a change in position of a body over a period of time). Kinematics describes the *spatial* (movement through space) and *temporal* (timing) aspects of movement; it isn't concerned with the forces that *cause* the movement. The kinematics of movement are the details of what you see when watching performance.

The basic spatial measures in kinematics include where, how far, how fast, and whether the body is speeding up or slowing down. The timing measures include when, how long, and the sequencing of the component parts of the movement, including the sequential joint actions.

Typecasting motion: Linear, angular, and general

The human body involves many segments, and both the body and the segments can move. All movement can be classified as either linear or angular:

✔ **Linear motion:** Linear motion occurs when all parts of a body undergo the same change in position during the same period of time. There are two types of linear motion (see Chapter 5):

• **Rectilinear motion** is motion in a straight line. All points on the body move exactly the same, and the orientation of the body does not change.

• **Curvilinear motion** is motion along a curved path. The body is moving simultaneously in two (or three) different planes. For example, when you throw a ball across the room, the ball follows a curvilinear path through the air. It moves up and then down at the same time that it moves away from you. This simultaneous movement in two different directions creates a parabolic path. The body can also be moving sideways along with the up–down and forward paths through the air, but it still follows a parabolic path. Parabolic motion of an airborne body is described in Chapter 5.

✔ **Angular motion:** Angular motion is the formal name for rotation — twists, spins, and turns. Rotation occurs when all points on the body go through the same angle. A point farther from the axis of rotation travels through the same angle as a point closer to the axis of rotation, but the curvilinear path traveled by a point farther from the axis of rotation is longer than the curvilinear path traveled by a point closer to the axis of rotation. The whole body can rotate, and the individual segments of the body rotate at the joints. (See Chapter 9 for more on rotation.)

General motion — a combination of angular motion and linear motion — is the most common form of motion. The linear motion of a point on a segment can only undergo linear motion because of a coordinated combination of angular motion at two or more joints. Synchronizing the angular motion of joints is a critical factor for successful performance, as explained in Chapter 16.

Describing how far: Distance and displacement

The terms *distance* and *displacement* are often used interchangeably in everyday conversation. Both terms describe how far a body moves. However, they have specific, and different, definitions in mechanics. Distance and displacement are measured in both linear and angular kinematics.

Distance

Distance is a scalar quantity. It simply describes how far the body moved during a time interval. Distance includes all the movement of the body from the start to the end of the motion, regardless of whether the body changes direction while traveling.

Linear distance is defined as the length of path traveled by a body, and it is measured in meters (m).

Angular distance is equal to the entire length of the angle traveled between the first and the final angular positions; it's measured in degrees (°). Some angles are measured in a unit called the *radian*, where 1 radian = 57.3 degrees. The steps to calculate linear distance are outlined in Chapter 5, and the steps to calculate angular distance are in Chapter 9.

Displacement

Displacement is a vector quantity. It describes the length of the change in position of a body in a specific direction during a time interval. Changes in direction during the time interval affect the measured displacement.

Linear displacement is measured as a straight line between where the body started and where the body ended up. Linear displacement is how far apart the start and end positions are — it's the shortest distance between the two points. The direction of the displacement must also be stated using a reference system, as explained in Chapter 5.

Angular displacement is the measured angle between the start and end positions of a rotating body. The direction of angular displacement is specified as clockwise or counterclockwise. The steps to calculate angular displacement are explained in Chapter 9.

Describing how fast: Speed and velocity

How fast a body moves refers to a change in position occurring over a period of time. The period of time refers to the difference between the time at the start of the interval and the time at the end of the interval over which the body is being evaluated. The time period can be long or short. Both how far the body moves and how long it takes to move affect the calculation of how fast. The term for how fast a body moves depends on whether how far is measured as a distance or as a displacement.

Speed

Speed is a scalar quantity. It simply describes how fast the body moved when it traveled the entire distance. You calculate speed by dividing the length of the path traveled by the length of time. In other words, $\text{speed} = \frac{\text{distance}}{\text{time}}$. In Chapter 5, I provide the steps you take to calculate linear speed (linear distance divided by time) in units of m/s; the steps to calculate angular speed (angular distance divided by time) in units of °/s are in Chapter 9.

Velocity

Velocity is a vector quantity. It describes how fast a body moved when it underwent a specific displacement. You calculate velocity by dividing displacement by the length of time. In other words, $\text{velocity} = \frac{\text{displacement}}{\text{time}}$.

Linear velocity, represented with v, is calculated as linear displacement divided by time (see Chapter 5), while angular velocity, represented with the Greek letter ω (omega) is calculated as angular displacement divided by time (see Chapter 9). To fully describe either v or ω, the direction of the body's motion must be specified.

Changing velocity: Acceleration

Acceleration measures the change in the velocity of a body over a period of time. Calculating acceleration considers the change in the velocity of the body relative to the length of the time period over which the velocity change occurred. Acceleration can be calculated for both linear motion (simply called *acceleration* and represented with the letter *a*) and angular motion (called *angular acceleration,* and represented with the Greek letter α [alpha]).

$$a = \frac{v_f - v_i}{t_f - t_i} \qquad \alpha = \frac{\omega_f - \omega_i}{\Delta t}$$

The subscripts *f* and *i* stand for "final" and "initial," respectively. For velocity, that means the velocity at the end of the interval and the velocity at the start of the interval; for time, the final time is the time recorded at the end of the interval and the initial time is the time recorded at the start of the interval. The difference between the final and the initial is called the *change in,* which can also be represented with the Greek symbol Δ (delta), as shown for the calculation of the change in time in the equation for angular acceleration.

The units for linear acceleration are m/s/s (meters per second per second); for angular acceleration, the units are °/s/s (degrees per second per second). The meters per second and degrees per second are the units of linear and angular velocity, respectively, and the last "per second" comes from dividing the change in velocity by the time period measured in seconds.

Acceleration is an important term in biomechanics because this quantity provides the link between kinematics (the motion we see) and kinetics (the forces causing the motion that we see). This link is explained for linear motion in Chapter 5 and for angular motion in Chapter 9.

Sometimes the term *acceleration* is used to refer to a body when it speeds up, and *deceleration* is used to refer to a body when it slows down. But the mechanical definition of *acceleration* is simply a change in velocity of the body, and the calculation is the same. The interpretation of a positive and a negative acceleration depends on the direction the body is traveling. I explain interpreting linear acceleration in Chapter 5 and angular acceleration in Chapter 9.

Pushing and Pulling into Kinetics

Kinetics is the branch of mechanics focused on the forces acting on a body. A *force* is a push or a pull created by the interaction of two bodies. Force is a vector quantity, and it's described by the characteristics of *magnitude* (the size of the force) and the direction of the force. Two other important characteristics of a force are its *point of application* (where on the body the force is applied) and its *line of action* (an imaginary extension of the force vector in both directions, used to determine the *torque,* or turning effect, of the force).

Inertia refers to a body's resistance to changing its motion. Essentially, inertia means that when a body is at rest, or not moving, it wants to stay at rest, and when a body is moving, it wants to continue to move at the same speed and in the same direction (the body tends to maintain its velocity). A good way to think of this is that when a body is changing its motion, a force, unopposed by another force, must be acting on it.

The inertia of a body comes from its mass. Mass is a measure of the quantity of matter in a body. The quantity of matter in a body refers to adding up the molecules of all the materials that make up a body. Fortunately, because there are a lot of molecules in a body and each molecule is really small, the mass of a body can be measured and reported as a single value without considering each individual molecule within it.

The pushes and pulls on a body — the forces — are classified as either internal or external:

- **Internal force:** A force produced within the body. An internal force doesn't affect the motion of the body — an internal force does not cause an acceleration of the whole body. An internal force pushes or pulls on the parts *within* the body, causing a deformation of the material called *strain,* depending on the magnitude of the stress produced. (Stress is a measure of how the applied force is distributed over the internal structure of the body; stress and strain are explained in Chapter 12.) When looking at the body as a whole, the pulling force of muscle is an internal force.

- **External force:** A force applied to a body by something outside the body. An external force acting on a body wants to change the motion of the body, to cause it to speed up or slow down (in other words, an external force wants to accelerate the body). We move around the environment because of the external forces that act on our bodies. We use our muscles to pull on our segments to produce external forces on our bodies.

    External forces fall into two categories:

    - **Non-contact force:** A non-contact force occurs when the two bodies are not in contact with each other. The most common external force in biomechanics is weight. The weight of a body is caused by the pulling force of gravity. The weight of a body is calculated as the mass of the body multiplied by the acceleration of gravity, or $W = mg$, where W is weight, m is the mass of the body (in kilograms), and g is the gravitational acceleration of –9.81 m/s/s. (For convenience, I usually use –10 m/s/s as the gravitational acceleration because this makes the calculation easier.) Weight is a downward-directed force on the body (as indicated by the minus sign). Weight and gravity are explained in Chapter 4.

It's important to remember that weight is a downward force. A very common error in working with mechanics is to overlook the downward force of a body's weight.

The weight of a body is always present (at least on earth). The weight of the body is the name given to the force of gravity acting on a body — *weight* and the *force of gravity* can be used as interchangeable terms. Because gravity pulls down on the body, weight always acts downward in the vertical direction. When a body is moving upward, the downward force of gravity acts to slow down its motion. When a body is moving downward, weight acts to speed up its motion. Although gravity is a non-contact force, it's ever present — whether the body is touching or not touching the earth, whether the body is submerged in water or not. The apparent "weightlessness" of a body in water is explained in Chapter 11.

- **Contact force:** A contact force occurs when two bodies touch. The force is created by the interaction of the two bodies. Sometimes the effect of the force is obvious, as when two bodies collide while moving, but a contact force is present between your feet and the ground even when you're standing upright and not moving. Friction is a contact force present when bodies tend to slide across each other, and friction always opposes the direction of the sliding. The factors affecting friction include the force squeezing the bodies together (called the *normal force*) and the materials in contact with each other where the sliding can occur (represented with the coefficient of friction). All things friction are explained in Chapter 4.

Forcing yourself to understand Newton's laws of motion

Sir Isaac Newton is famous for his three laws describing the effect of force on a body. The laws are important because they fully describe the cause–effect relationship between force and motion: Force causes motion. Newton's three laws are the backbone of biomechanics. They provide the foundation for movement analysis.

Here's a very brief overview of the three laws (I cover them in much greater detail in Chapter 6):

✔ **Newton's first law — the law of inertia:** This law basically states that a body can only speed up or slow down if an external force acts on the body. Most important for biomechanics is to accept the fact that an unopposed force must act on the body to speed it up or slow it down.

✔ **Newton's third law — the law of action–reaction:** Here are the two important things to remember about Newton's third law:

   • The magnitude of the force on each body is equal in magnitude to the force on the other body.

   • The direction of the force on each body is opposite to the direction of the force on the other body.

The forces on the two bodies are equal and opposite regardless of the difference in the size and the mass of the two bodies.

For the scoop on why I've listed Newton's third law before Newton's second law, turn to Chapter 6.

✔ **Newton's second law — the law of acceleration:** The law of acceleration expresses the cause-and-effect relationship between force and motion. The law is usually stated as $F = ma$, where F is the unbalanced force, m is the mass of the body the force acts on, and a is the acceleration of the body caused by the unbalanced force.

The size of the acceleration is directly proportional to the size of the unbalanced force:

   • The size of the acceleration is larger if the force is larger.

   • The size of the acceleration is smaller if the force is smaller.

The size of the acceleration is inversely proportional to the mass of the body:

   • The size of the acceleration is larger if the mass is smaller.

   • The size of the acceleration is smaller if the mass is larger.

Falling apples change our view of the world

Many people who don't know Newton's first law from Newton's third law have heard the story of Newton and the apple. Sir Isaac Newton was 23 years old and a student at Cambridge University when the plague struck London in 1665. To avoid the disease, Newton went to his family home in the country to study and work on his ideas about force and motion. An apple falling in the garden helped him to draw together his ideas and to formulate the law of gravitation along with his three famous laws of motion. These theories were published in 1667 in his master work titled *The Principia*. Our study of the universe, including human movement, has never been the same.

Using the impulse–momentum relationship

Momentum is the product of a body's mass and velocity (momentum = mass × velocity). Momentum is a single quantity combining the body's inertia, or resistance to changing motion (mass) with how fast and in what direction it's traveling at an instant in time (its velocity). Any moving body has momentum.

All movement can be considered in terms of the desired momentum of a body. In a performance, we want a body, which has a mass, to be moving at a specific velocity. That is, the body has a specific momentum for the successful outcome of the performance. The faster the body is moving, the greater its velocity, the more momentum it has. A body at rest has no momentum, because its velocity is zero.

The mass of the body, in most cases in biomechanics, stays constant during a performance. The momentum of a body changes because of a change in velocity. Starting a body from rest, or making it go faster, increases the body's momentum. Conversely, bringing a body to rest, or making it go slower, decreases the body's momentum.

Acceleration refers to an increase or a decrease in velocity and is caused by an unbalanced force. In Chapter 6, I show how substituting the equation for acceleration as a change in velocity over time into the equation for Newton's second law leads to the impulse momentum relationship, or $Ft = m(v_f - v_i)$.

Ft is a quantity known as *impulse*. It represents the product of a force applied for a period of time. In Chapter 16, I explain how the mechanical objective of a performance is to apply the required impulse to the body to give it the momentum needed for success — the force applied for a period of time is what the performer can control, and using the impulse–momentum relationship is an objective way to evaluate performance.

Working with Energy and Power

Another way to look at movement is using the terms *mechanical work, mechanical energy,* and *mechanical power* (see Chapter 7).

Mechanical work

Mechanical work is defined as the product of force and displacement. When a force is applied to a body and the body moves or is moving, mechanical work is performed. In this book, when I refer to work, I'm referring to mechanical work. There are two types of mechanical work:

- **Positive work** is performed when the body moves in the same direction as the applied force.
- **Negative work** is performed when the body moves in the opposite direction as the applied force.

No mechanical work is performed when a force is applied to a body but the body doesn't move.

Mechanical energy

Mechanical energy is defined as the capacity of a body to do work. Any type of energy can be converted from one form to another form, and in the conversion, work is performed. In biomechanics, we're primarily concerned with *mechanical energy*, the energy a body has because it's moving, because of its position, or because it has been deformed.

Kinetic energy (KE) is the energy present in a body when it's moving. The amount of KE is equal to one-half of the product of the mass of the body and its velocity squared, or $KE = \frac{1}{2}mv^2$.

Gravitational potential energy (GPE) is the energy present in a body because of its position above a reference point. The amount of GPE is equal to the product of the mass of the body, the gravitational acceleration (g), and the height it will fall to the reference point (h), or $GPE = mgh$.

Strain energy (SE) is the energy present in a body when it's deformed. The amount of SE in a body is equal to one-half of the body's stiffness (k, a measure of its resistance to deformation) multiplied by the amount of deformation (Δx) squared, or $SE = \frac{1}{2}k\Delta x^2$.

The energy in a body can perform work — that is, the energy can produce a force on a body in the following conditions:

- The KE in a body can perform work on another body that it contacts.
- The GPE in a body can perform work because it will pick up downward velocity as it falls, and the potential energy, because of its position, will convert to KE.
- The SE in a body can perform work on the body as it returns to its original shape.

Mechanical power

Mechanical power is defined as the rate of doing work, or power $= \frac{\text{work}}{\text{time}}$. Greater power is produced by doing more mechanical work in the same period of time, or by producing the same mechanical work in a shorter period of time.

Turning Force into Torque

The turning effect of a force, its tendency to cause a body to rotate, is called *torque*. Torque is produced when an external force is applied to a body and the line of action of the force doesn't pass through an axis around which the body is able to rotate.

The magnitude of the turning effect depends on the magnitude and direction of the force and the moment arm (MA). As an equation, that's torque = force × moment arm, or $T = F \times MA$. The moment arm measures the perpendicular distance from the line of action of the force to an axis of rotation. A longer MA increases the turning effect of the force. Torque and moment arm are both explained in Chapter 8.

The body's resistance rotation is called the *moment of inertia* (I). The moment of inertia depends not only on the mass of the body, but also on how the mass is distributed around the axis of rotation. In fact, the distribution of the mass has a much greater effect than the size of the mass, as evident in the equation for moment of inertia, $I = mr^2$. In Chapter 10, I explain how a performer can manipulate the moment of inertia of the body to affect its rotation.

Dealing with Measurement Units

Things get measured in appropriate units. The easiest units to use come from the International System of Units (SI), a standardized set of units based on the metric system. It's easiest because the metric system is based on multiples of ten. The SI is used in most parts of the world, and almost exclusively in sciences like biomechanics.

The base units are those for length, time, and mass. All other units are a combination of these units, depending on what's being measured. When different units are combined, the new unit can be almost intimidating. Force, for example, is a combination of mass, length, and time, and the unit is kg · m/s/s. For ease or reporting, and in honor of Sir Isaac, the unit of force is called the Newton (1 N = 1 kg · m/s/s).

Table 3-1 presents the base units in the SI for length, time, and mass. It also shows other derived units for quantities commonly measured in biomechanics.

Table 3-1	Basic Units in Biomechanics	
Measurement	*Units*	*Abbreviation*
Length	Meters	m
Time	Seconds	s
Mass	Kilograms	kg
Force	Newtons	N
Angle	Degrees or radians	° or rad
Linear speed and velocity	Meters per second	m/s
Angular speed and velocity	Degrees per second	°/s
Linear acceleration	Meters per second per second	m/s/s or m/s^2
Angular acceleration	Degrees per second per second or radians per second per second	°/s/s or rad/s/s
Momentum	Kilogram meters per second	kg · m/s
Impulse	Newton seconds	N · s
Work and energy	Joules (Nm)	J
Power	Watts (Nm/s)	W
Torque or moment of force	Newton meters	Nm
Moment of inertia (I)	Kilogram meters^2	kg · m^2
Normal stress	Newtons per meter^2	σ
Shear stress	Newtons per meter^2	τ

Using the Neuromusculoskeletal System to Move

The "bio" in biomechanics refers to a living body. The human body is the focus of the biomechanics in this book, and the main system of the human body affecting movement is the neuromusculoskeletal (NMS) system, which actually consists of three systems. The nervous (neuro), muscular (musculo), and skeletal systems are covered in separate chapters.

The skeletal system

The skeletal system consists of the bones, joints, and ligaments of the body. These specialized structures provide support and protection for the body while still allowing for the individual body segments to move.

Bones

Bone provides structure and protection for the body. Bones are classified according to their shape and size. Bone is a dynamic tissue that changes during growth and development and in response to imposed stress. Bone also serves as a storage site for many minerals used by the body, the most important of these being calcium. Calcium provides the strength for bone, and a loss of calcium reduces the strength of the bone and makes it easier for a fracture to occur. This is a condition called *osteoporosis* (see Chapter 13).

Joints

A *joint* is a site where bones meet. Although the bones meeting at some joints are fused together to provide protection (the bones of the skull for example), most joints of the body allow movement between the bones. The most important joint for movement is the *synovial joint,* where the ends of the bones are covered in cartilage to allow easy sliding. A synovial joint also has a joint capsule made of tough connective tissue to help hold the bones together at the joint. The interior of the synovial joint is filled with synovial fluid, with the consistency of egg white, to lubricate and nourish the cartilage. With healthy cartilage and synovial fluid, the synovial joint provides almost no resistance to the movement of the bones. The synovial joint is explained in more detail in Chapter 13.

Ligaments

A *ligament* is a specialized connective tissue providing stability to a joint. Most ligaments are part of the joint capsule, located to reinforce the capsule where the joint is most susceptible to *dislocation,* or coming apart. In spite of its great strength, a ligament can be stretched by loading to the point of injury, called a *ligament sprain.* Sprains are one of the most common injuries related to physical activity. There are different degrees of ligament strain, from Type I (mild) to Type III (severe, because it involves a tear of the ligament). Differences in the degree of ligament sprain are why the recovery time from some "sprained ligaments" is quite short while the recovery time from other "sprained ligaments" is very long, possibly involving surgery and a long, challenging rehabilitation.

The muscular system

The muscular system of the body consists of the individual muscles and the tendons, which attach the muscles to the bone.

Muscles

Muscle is a specialized tissue with the unique capability to produce force. Muscle comes in different forms throughout the body:

- ✔ Cardiac muscle is found in the heart and arteries of the body. They pump blood.

- ✔ Smooth muscle lines the digestive system and pushes food along and out of the body.

- ✔ Skeletal muscle is attached to bones by tendons, and the muscle and tendon act together to start, stop, and control the movement of the segments.

The word *muscle* is used to refer to an individual identifiable muscle, like your biceps or triceps muscle or your hamstring muscle (a group of three muscles on the back of your hip and knee). The term *muscle* is also used to refer to the specialized cells found within a given muscle. Muscle cells are called *fibers* and consist of a long chain of individual tension-producing units called *sarcomeres*. The sarcomeres contain thin protein filaments called *actin* and *myosin,* and the interaction between actin and myosin produces the muscle force, or muscle tension. This force is also known as *active tension* because it's created by the interaction of the actin and myosin. Chapter 15 presents more detail on the structure of muscle and how it produces force.

The one most important, and often overlooked, aspect of all muscle is that a muscle can only pull. Muscles can't push. The pull of a muscle is called *muscle tension.* Because muscles can only pull, they're arranged in groups around a joint to produce and control the motion of the segments meeting at a joint. Different muscle groups tend to cause different joint actions, but all muscles produce force by pulling.

There are three types of muscle tension development:

- ✔ **Isometric:** The muscle produces tension, but the segments it pulls on don't move. The muscle stays the same length.

- ✔ **Concentric:** The muscle gets shorter while it produces tension. In this type of muscle activity, the muscle tension pulls the attachment sites closer together. Concentric activity causes the observed joint motion.

- ✔ **Eccentric:** The muscle gets longer while it produces tension. In this type of muscle activity, the muscle tension pulls, but the attachment sites move farther apart. Concentric activity resists the observed joint motion.

Tendons

Tendons attach bones to muscles. A tendon can be shaped like a rope (feel the tendon on the back, or *posterior,* of your ankle), or it can be broad and thin (like the tendons attaching the muscles on the front, or *anterior,* of your thigh to the top, or *superior,* surface of your knee cap).

The material making up a tendon is tough and *elastic,* meaning that it resists tearing, but it can be stretched and then return to its original length. Making use of the elasticity of the tendon is important to effective movement. Using the elasticity of the tendon during movement performance is part of the stretch-shorten cycle. The stretch-shorten cycle consists of actively lengthening a muscle *(eccentric activity)* and then, without any hesitation or delay, actively shortening the same muscle *(concentric activity).* The stretch-shorten cycle increases the force produced by the muscle because the tension produced is the sum of the active tension from the muscle and the passive tension from the tendon. I explain the passive tension from the tendon in Chapter 15 and show the importance of the stretch-shorten cycle to performance in Chapter 16.

The nervous system

The nervous system is the communication and control system of the body. The fundamental unit of the nervous system is the *neuron,* or nerve cell. The nervous system consists of the brain, the spinal cord (the central nervous system, or CNS), and the peripheral nervous system (PNS), consisting of all the nerves carrying signals to the muscles (motor nerves) and from the muscles (sensory nerves), organs, and glands of the body.

Like all systems of the body, the nervous system is well organized with recognizable features. In this book, I focus on the most critical aspects of the nervous system for controlling movement, without going in depth into the cognitive aspects and structures of the brain. Look in Chapter 16 for more detail on the nervous system.

Motor nerves

Motor nerves communicate signals from the spinal cord to the muscles. The signal is called an *action potential,* and it consists of a low-voltage electrical impulse. A single motor nerve activates multiple muscle fibers. Activating more motor nerves activates more muscle fibers and provides a means to increase the amount of tension, or *muscle force,* produced by a muscle.

Sensory nerves and receptors

Sensory receptors are located throughout the body to detect certain kinds of stimuli. When a sensory receptor is activated, it sends an action potential back to the CNS along a sensory neuron. The sensory neuron meets with (or *synapses*) other neurons, and a response, if any, is determined.

The sensory receptors discussed in this book are the muscle spindle and the Golgi tendon organ.

Muscle spindles are located within the muscles. These sensory receptors detect changes in muscle length and the rate of the change in muscle length. When a muscle spindle is stimulated, the response is to cause the muscle to be activated. The muscle spindle response is called an *excitatory response,* because it activates the muscle that caused it to respond.

Golgi tendon organs are located where the muscle fibers join with the connective tissue of the tendon. The Golgi tendon organ detects the amount of tension, or pulling force, produced by the muscle. When the tendon organ is activated, the response is to decrease the activation of the muscle. The muscle spindle response is called an *inhibitory response,* because it decreases activation in the muscle that caused it to respond.

Part II
Looking At Linear Mechanics

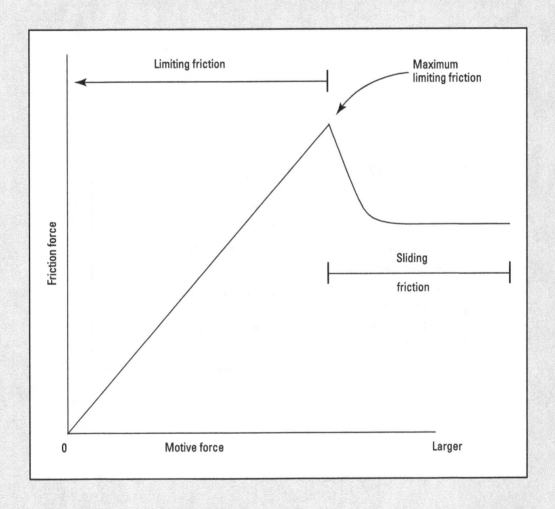

Dig into composing a resultant force vector from multiple vectors in an article at
http://www.dummies.com/extras/biomechanics.

In this part...

- Use Newton's Laws to show how force causes motion.

- Get the lowdown on gravity and friction.

- Work with linear kinematics and parabolic motion.

- Calculate the net force on a body.

- Apply Newton's laws through the impulse–momentum relationship.

- Discover the relationships among work, energy, and power.

Chapter 4

Making Motion Change: Force

*Y*ou hear the word *force* a lot in everyday language. Kids claim they were forced to do homework. The force of your argument can convince me your claim is correct. And in a galaxy far, far, away, people bid a fond farewell with the statement "May the force be with you." In mechanics, however, *force* has a precise meaning, and sticking to that meaning makes understanding mechanics a lot easier.

I begin this chapter by defining the mechanical concept of force. I show you how to work with force using the tools of the right-angle triangle. Then I walk you through the classification of force — as contact or noncontact, internal or external. Finally, I define the important forces of gravity and friction. This chapter gives you the basic tools you need to understand how force and motion are related.

Pushing and Pulling: What Is Force?

Put simply, a *force* is a push or a pull. Muscles pull on bones to affect their motion. You can pull on a handle to lift up a suitcase. To walk, your muscles pull on the bones of your leg, and the moving segments of your leg result in a push on the ground. In soccer, when your foot (or head!) strikes a ball, the ball gets a push. When you collide with someone on a sidewalk, the collision applies a push to both of you.

Although you can often see a facial grimace on a power lifter telling us he or she is exerting a great amount of force, technically, forces are not visible. A force is inferred by the observed changes in motion (or lack of changes in motion) of the body on which the force acts. (The *body* in biomechanics is the specific object being analyzed — not necessarily a human body.) Sometimes the effect itself is not all that evident unless sensitive instruments are used. But just because you can't *see* a force, doesn't mean a force doesn't exist.

Every force has the following characteristics:

- ✔ **Magnitude:** Size; how big the push or pull is

- ✔ **Direction:** Which way the force is pushing or pulling

- ✔ **Point of application:** Where the force is acting on the body

- ✔ **Line of action:** An imaginary extension of the force in both directions, which is useful in determining the turning effect of a force

Force is a vector quantity, with both magnitude and direction. To describe a force, you must specify both — one descriptor without the other leaves the force undefined. Because force is a vector quantity, you represent it graphically with an arrow. The length of the arrow represents the force magnitude; the longer the arrow, the bigger the force. The tip of the arrow identifies force direction. Typically, a standard Cartesian coordinate system is used to represent the direction of a force, with the *y*-axis the vertical direction and the *x*-axis the horizontal direction. The positive directions are upward (+, vertical direction) or to the right (+, horizontal direction), and negative directions are downward (–, vertical direction) or to the left (–, horizontal direction).

Figure 4-1 shows a few graphical examples of force drawn on a coordinate system. The forces are named F_1 through F_8. Note how some forces are similar in magnitude but have different directions (F_1 and F_2), other forces are different in magnitude but have the same direction (F_3 and F_4), and other forces differ in both magnitude and direction (F_5 and F_6).

Force is measured in Newtons in the International System of Units (SI) and in pounds in the English system. One Newton of force is equivalent to 4.45 pounds of force. So, if you weigh 100 pounds, you weigh 445 Newtons. The value of Newtons or pounds represents the size, or magnitude, of a force. One Newton of force is defined as the force required to accelerate a 1 kg mass at the rate of 1 m per second per second (m/s/s), or mathematically $1\,N = 1\,kg \cdot 1\,m/s/s$. (This comes from Newton's equation $F = ma$, covered in detail in Chapter 6.)

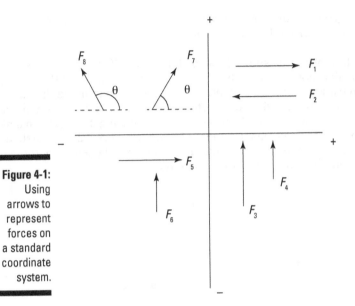

Figure 4-1:
Using
arrows to
represent
forces on
a standard
coordinate
system.

A push of 30 Newtons to the right is described as +30 N in the horizontal direction, while a push of 30 Newtons to the left is described as –30 N in the horizontal direction. Similarly, a force of 200 Newtons upward is described as +200 N in the vertical direction, and a push of 200 Newtons downward is described as –200 N in the vertical direction.

When a force is applied at an angle, the angle of application is provided relative to the right horizontal axis, or a specified reference line. The angle of force application is represented with the Greek letter θ, or theta. Figure 4-1 shows forces F_7 and F_8 applied at an angle.

Specifying the magnitude and the direction of a force allows you to visualize the force with a vector (arrow).

Magnitude and direction are two critical characteristics of a force, but two other characteristics — point of application and line of action — are also very important.

The point of application indicates where the force is applied on a body. It's represented by drawing the force touching the body at — you guessed it — the point of application. Either the tip or the tail of the force can be indicated at the point of application. The point of application is a simplification of how a force is applied to the body. For example, when you stand, the force of the ground pushing up on you is applied over the soles of your feet where your feet touch the ground. But you can draw one vector under each foot to represent the force of the ground under each foot.

The line of action of a force shows how the force also affects (or can also affect) the angular motion of a body. The line of action influences the turning effect of a force and is covered in depth in Chapter 8. You indicate the line of action of a force by extending the vector indefinitely in both directions. A good way to represent the line of action without confusing it with the magnitude of the force is to use a dashed line, as shown in Figure 4-2. The dashed line is an extension of the vector in both directions, but the length of the dashed line is not an indicator of force magnitude.

Figure 4-2:
Force vectors with the line of action represented by the dashed line.

Force vector

Line of action of the force

The line of action of a force is *not* the same thing as the direction of a force. They're both important, but for different reasons.

To describe a force, you must specify at least the magnitude and the direction. To *fully* describe a force acting on a body, and to understand the effect a force will have on the body, you must specify the magnitude, direction, point of application, and line of action. Figures 4-3 through 4-5 provide examples of how the characteristics of a force are represented when drawing forces on a body.

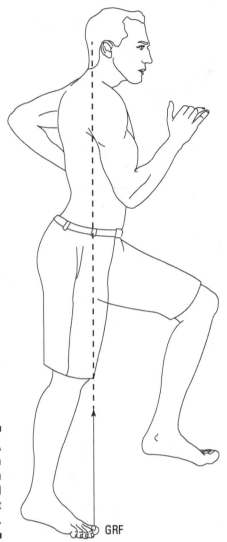

Figure 4-3: A runner with a force from the ground applied at the foot.

GRF

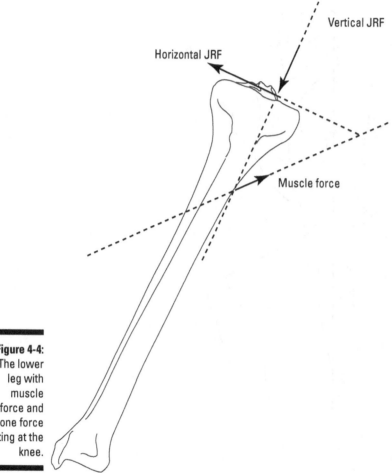

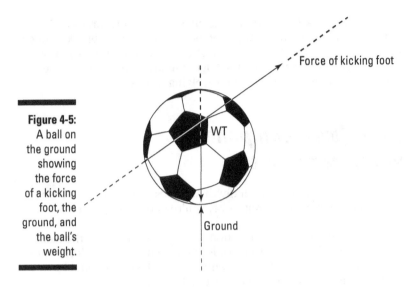

Figure 4-5:
A ball on
the ground
showing
the force
of a kicking
foot, the
ground, and
the ball's
weight.

Working with Force Vectors

When multiple forces act on a body, each force tends to affect the body. The net force on the body reflects a contribution from the individual forces. Lay this book on a table. Put your right hand on the bottom-right corner of the book, and push on the book toward the top-left corner. The book slides at an angle, moving in the direction of the force you apply. Note that the book slides both forward and to the side simultaneously, although your force was applied at an angle.

Similarly, two forces applied to a body can affect the body as if one force were applied. With this book on the table, put your left hand in the bottom-center of the book and your right hand on the center of the right side of the book. Push forward with your left hand, and push to the left with your right hand. Note how the book slides at an angle, similar to when you applied a single push on the book at an angle with your right hand. Neither of your forces was applied on this angle, but that is the direction the body moved. This simple exercise shows two things about forces:

✔ Two forces, added together, can have the same effect as a single force.

✔ A single force applied alone can have the same effect as two forces acting together.

In this section, I show how to determine the effect of multiple forces acting on a body at the same time (composition of forces) and how to break a single force into equivalent forces acting at right angles to each other (resolution of force). The Pythagorean theorem and the trigonometric functions, covered in Chapter 2, are used when composing and resolving forces.

Using the force components to find the resultant

When you add together two or more individual forces (components) acting on a body to create a single, resultant force, you're composing a force.

Colinear forces share the same line of action. The forces can act in the same or opposite directions. Composing a force is simple when the two forces are colinear. If horizontal forces F_1 of +200 N and F_2 of –100 N are applied to a body, the net force is simply the sum of the two forces:

$$\text{Net Force } (\Sigma F) = F_1 + F_2$$
$$\Sigma F = (+200) + (-100)$$
$$= +100 \text{ N}$$

Because forces are vectors, always include the direction sign when referring to a force. It's a good idea to use parentheses to separate the forces when listing them in an equation. Within the parentheses, identify the direction of the force as + or –. This simplifies adding the forces and decreases the chance of making an error when solving for the resultant force.

Now, consider the two forces acting on the figure skater in Figure 4-6: a vertical force F_1 of +600 N and a horizontal force F_2 of +200 N. You can't simply add the two forces for a resultant force of +800 N on the skater because this ignores the different *directions* of the forces. Instead, the resultant force is calculated using the Pythagorean theorem for the magnitude and the arctan (inverse tangent) for the angle (see Chapter 2).

Use the Pythagorean theorem and the arctan trig function to compose the resultant force from two or more forces in the vertical and horizontal directions.

Make sure your calculator is in the correct mode of degrees, and not in radians, when working with the trig function (unless the angle is measured in radians, covered in Chapter 9).

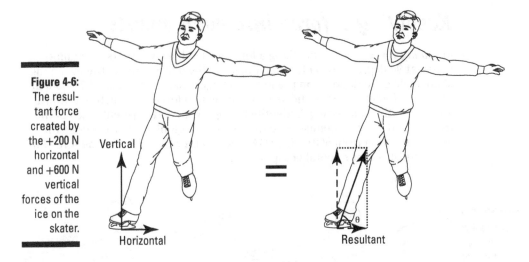

Figure 4-6:
The resul-
tant force
created by
the +200 N
horizontal
and +600 N
vertical
forces of the
ice on the
skater.

Visualize the two forces — horizontal and vertical — as creating a single force on the leg of the skater (refer to Figure 4-6). Drawing in lines parallel to the two forces creates a parallelogram, with the diagonal representing the resultant force. The angle of the diagonal from the original horizontal force is the direction of the resultant force. The right triangle created allows use of the Pythagorean theorem to solve for the magnitude and the arctan to solve for the direction of the resultant force.

The magnitude of the hypotenuse is solved as follows:

$$\text{Resultant}^2 = \text{Vertical}^2 + \text{Horizontal}^2$$
$$\text{Resultant} = \sqrt{\text{Vertical}^2 + \text{Horizontal}^2}$$
$$= \sqrt{600^2 + 200^2}$$
$$= \sqrt{360,000 + 40,000}$$
$$= \sqrt{400,000}$$
$$= 632.5\,\text{N}$$

The direction of the resultant force is solved using the arctan function:

$$\theta = \arctan \frac{\text{Vertical}}{\text{Horizontal}}$$
$$= \arctan \frac{600}{200}$$
$$= \arctan 3$$
$$= 71.6°$$

The resultant force is 632.5 N at an angle of 71.6°.

Resolving a force into components

Resolving a force means breaking a force vector into two parts, or components, at 90 degrees to each other. Consider the 920 N force of the ice acting at an angle of 70 degrees on the leg of the figure skater in Figure 4-7. Visualize the single force as having the same effect as two forces, one acting upward and one to the side. Using θ, identify the opposite and adjacent sides of the triangle and the hypotenuse. Use the process of elimination to choose the correct function from SOH CAH TOA (see Chapter 2) to resolve the force into the vertical and horizontal components.

Figure 4-7:
Resolve a force of 920 N at an angle of 70 degrees into the vertical and horizontal components using the sine and cosine.

Recognize the horizontal force as the side *adjacent* to the 70-degree angle of the right triangle, and create a table of variables:

$\theta = 70°$

Hypotenuse = 920 N

Adjacent (Horizontal Force) = ? N (the unknown, in Newtons)

Using the process of elimination, select the cosine from SOH CAH TOA:

$$\cos\theta = \frac{\text{Adjacent}}{\text{Hypotenuse}}$$

Isolate for the unknown side, the horizontal side, and solve:

Adjacent = cos θ · Hypotenuse = cos 70° · 920 = 0.342 · 920 = 314.7 N

The horizontal component of the ice force on the skater is 314.7 N.

Now, solve for the vertical component. Recognize the vertical component as the side *opposite* the right angle, and create a table of variables:

$\theta = 70°$

Hypotenuse = 920 N

Opposite (vertical force) = ? N (the unknown, in Newtons)

Select the sine as the appropriate trig function from SOH CAH TOA:

$$\sin \theta = \frac{\text{Opposite}}{\text{Hypotenuse}}$$

Isolate for the unknown side, the opposite side, and solve:

Opposite = $\sin \theta \cdot$ Hypotenuse = $\sin 70° \cdot 920 = 940 \cdot 920 = 864.5$ N

The vertical component of the ice force on the skater is 864.5 N.

Make sure to assign Opposite and Adjacent to the correct sides of the right triangle. If you reverse them, you'll still correctly calculate the component magnitudes, but you won't assign them to the correct component.

Apply the Pythagorean theorem to check your calculation of the magnitudes of the two resolved sides, showing that 920 is, indeed, $\sqrt{314.7^2 + 864.5^2}$.

Classifying Forces

A force is a push or pull, and arrows (vectors) represent the magnitude, direction, point of application, and line of action. But where does a force come from? How will it affect the body? The source of a force leads to the classification of a force as either a contact or a noncontact force, and its ability to affect the motion leads to the classification as either an internal or an external force acting on the body. If you keep in mind that not much creativity was used to classify forces, you're ahead of the game.

Contact and noncontact forces

Most forces come when bodies touch each other, but one of the most important forces occurs whether the bodies are touching or not. I describe contact and noncontact forces in this section.

Contact forces

A *contact force* occurs when — drumroll, please — two bodies are in contact with each other. (Amazing, I know.) Here are some examples of contact forces:

- The force of your hand on this book
- The force of your butt on the chair
- The force between your foot and the ground when walking
- The force of a bat striking a ball
- The force of a muscle pulling on a bone

In each of these examples, *two* bodies are identified (hand/book; butt/chair; foot/ground; bat/ball; muscle/bone), and the bodies contact each other. Thus, these are contact forces.

Contact forces can be of varying size and varying direction, with a varying point of application and line of action.

Noncontact forces

If contact forces occur when two bodies are in contact with each other, then noncontact forces occur when — eureka! — the bodies are *not* in contact with each other. (If all mechanics were this easy, you wouldn't need this book.)

The important noncontact force in biomechanics comes from gravity, the downward pull of the Earth on all objects. Gravity is present even when a body is not touching the Earth. Jump into the air — you're not touching the Earth, but the Earth still pulls down on you. The noncontact gravitational force slows you down and stops you as you rise into the air, and it speeds you up as you fall back toward the ground (where a contact force between you and the ground will stop your downward motion — just try to land on your feet!). The same pull occurs when a body like a ball is thrown into the air: Gravity slows the ball down on the way up and speeds it up on the way back down. Another name for gravitational force is *weight* (see "Feeling the Pull of Gravity," later in this chapter).

Internal and external forces

Why can't a cat reach around, pull up on its tail, and lift itself into the air? For that matter, why can't you pull yourself out of your chair by pulling up on your hair? Forces, in the form of pulls, are obviously present (the cat pulling on its tail, you pulling on your hair), so why doesn't the cat or your body rise up? The answer has to do with the classification of a force as external or internal.

Internal forces

An *internal force* acts within, or internal to, the body being analyzed. An internal force doesn't change the motion of the body being analyzed. Consider the cat as a whole. Ignoring the fact that cats don't have opposable thumbs or even hands, the cat's paw and its tail are both part of, or internal to, the cat itself. So, the pull of the cat's paw on the tail and the pull of the tail on the hand are both *internal* to the cat. The two forces act on different parts of the same body. The two pulls cancel each other out when considered relative to the whole body under analysis. The forces can't, and don't, affect the motion of the body as a whole.

Internal forces are important because they reorient segments in the body and contribute to the stability of the body. Internal forces become important when looking at the effect of physical activity on training and injury of tissues, as explained in Chapter 12.

External forces

An *external force* acts from outside, or external to, the body under analysis. An external force can cause a change in the motion of the body. Focus on the tail (or, go ahead, pull your hair again). With the tail as the focus of the analysis, the pull of the cat's paw is external to the tail. The pull of the paw causes the tail to rise, a change in motion.

External forces are important because they cause, or tend to cause, a change in motion of the body acted on. Identifying the external forces acting on a body is a critical step when looking at how bodies change motion.

Two bodies interact to create the force. If the two bodies are within the specific body under analysis, they're internal forces. If one of the two bodies is your focus of analysis, the force on it from the other body is an external force. Go ahead — pull your hair and think about it.

A free-body diagram (FBD) is a sketch of the body under investigation showing all the external forces. (Note that I call it a *sketch,* not a detailed drawing.) Creating an FBD forces you (pun intended) to identify all the external forces acting on a body.

A sequence of four steps makes it easier to create an FBD. Let's draw an FBD of someone weighing 624 N (or 140 pounds) standing on the floor on one leg and leaning against the wall (see Figure 4-8).

Figure 4-8:
A guy
standing
on one leg
and leaning
against the
wall with his
hand.

1. **Precisely specify the body under investigation, and draw a simplified figure to represent the body.**

   In this case, the body is the entire person. So, you draw the body as a rectangle with stick-like arms and legs. Ignore the detail of the body parts. Sketch only the body under investigation, isolated from everything else — no floor, no wall.

2. **List all the external forces acting on the system.**

   This step is easiest if you identify, in order, the noncontact forces and then the contact forces:

   - **Noncontact forces:** The only external noncontact force is the body weight, from gravity.

     Unless you're studying someone in outer space, where gravity is missing, *always* include body weight. Gravity is the only noncontact force to consider, and we have to consider it all the time.

   - **Contact forces:** What is touching the body? The person being analyzed is standing on the ground, so obviously there is a contact force from the ground. The person is also touching the wall, so there is a contact force from the wall.

   The three external forces — weight, ground force, and wall force — must be drawn on the FBD.

3. **Consider the magnitude, direction, and point of application of each of the external forces.**

   Taking the three external forces one by one, you have the following:

   - **Weight:** The magnitude is 624 N, the direction is downward, and the point of application is at the person's *center of gravity* (the point of application of the weight of a body).

     Although the center of gravity differs slightly in men and women, it's generally at about the height of the belly button. (You can find more detail on the center of gravity in Chapter 8.)

   - **Ground force:** The magnitude and direction are unknown (given the information we have in this situation), and the point of application is at the person's foot.

   - **Wall force:** The magnitude and direction are unknown (given the information we have in this situation), and the point of application is at the person's hand.

4. **Sketch vectors representing the forces on the body.**

   Your end result should be something similar to what you see in Figure 4-9. All the external forces are represented on the body.

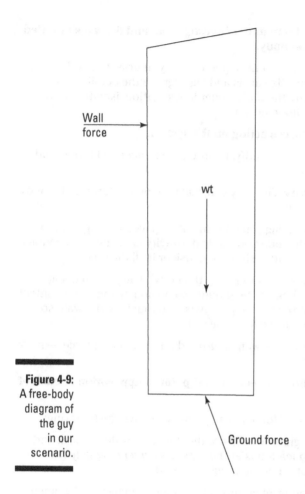

Wall
force

wt

Figure 4-9:
A free-body
diagram of
the guy
in our
scenario.

Ground force

Your FBD doesn't have to show each force accurately — creating an FBD requires you to think through the identification of the external forces acting on the body. The technique to solve for the magnitude and direction of the forces identified in the FBD is shown in Chapter 6.

Feeling the Pull of Gravity

Weight is the force of the Earth's gravity pulling on a body. Weight is a down-ward-acting, external force that causes a body to slow down as it rises in the air and to speed up as it falls back toward the Earth. We're so accustomed to dealing with our own weight — working against it as we climb stairs, jump into the air, or simply stand in place — that it's easy to overlook this ever-present force when working in mechanics. But gravity influences motion more than almost any other force.

Newton explained gravity in his law of gravitation, which in three sentences essentially states:

- All bodies pull toward each other with a force.

- The magnitude of the pulling force is proportional to the product of the mass of the two bodies.

- The magnitude of the force is inversely proportional to the square of the distance between the two bodies.

Whew! Here's how each of these statements applies to mechanics:

- **All bodies pull toward each other with a force.** With this sentence, Newton sets the stage for gravity. Everything in the universe pulls on everything else. The planets and the sun pull on each other. You're being pulled by every object in the universe (although it may not seem like that unless it's exam time and you still haven't finished your holiday shopping). But you aren't being tugged around all the time, in all directions, every which way. Fortunately, almost all these pulling forces are cancelled by other pulling forces, so these innumerous pulling forces can be ignored. But *one* pulling force cannot be ignored: the pull of the Earth. This pulling force is gravity, and the pulling force of gravity is commonly called weight. Weight is the pulling force of the Earth on a body. Because gravity is a pulling force, weight acts toward the surface (actually the center) of the Earth. Weight is a force vertically downward, represented by the letter W, for weight.

- **The magnitude of the pulling force is proportional to the product of the mass of the two bodies.** *Mass* is the amount of matter in a body (see Chapter 3). The two masses you consider in order to calculate weight are the Earth and some other body (for example, you). The Earth has a lot of mass (take my word for it — although estimates of the Earth's mass vary depending on the source, in every case it is a *very* large value). You, on the other hand, have a lot less mass than the Earth does. So, the relatively large mass of the Earth pulls on your relatively small mass, while at the same time, your relatively small mass pulls on the relatively large mass of the Earth.

  Let's call the mass of the earth m_1 and the other mass (your mass) m_2. So the product of these two masses is $m_1 m_2$.

- **The magnitude of the pulling force is inversely proportional to the square of the distance between the two bodies.** The distance to consider is from the surface of the Earth to the center of the Earth. Because no one has bored through to the center of the Earth, this distance is estimated, and estimates vary. What is consistent is that the distance is long — so long that, for our purposes, we can ignore the fact that the surface of the Earth varies in distance from the center of the Earth, depending on where you are. (In other words, the distance to the center of the Earth is greater at Mount Everest than it is in the Dead Sea, but you can ignore this fact.)

Let's call the distance from the center of the Earth to your body on the surface of the Earth r (for the radius of the Earth). So, the square of this distance is r^2. Newton's equation for the force of gravity on you is

$$W = \frac{m_1 m_2}{r^2}$$

As $m_1 m_2$ increases, so does the weight — weight is proportional to the product of the masses. As r^2 increases, the weight decreases — weight is inversely proportional to the square of the distance between the two bodies.

In the equation, m_1 and r^2 are constant for any and all calculations of the weight of any body on the surface of the Earth. The mass of the Earth does not change, nor does the distance to the center of the Earth. The only term on the right side of the equation that changes is m_2, the mass of the body being analyzed. So, weight is proportional to m_2 (which is written $W \sim m_2$). The Earth has a greater pulling force on a body with more mass than it does on a body with less mass. That is, a body with more mass weighs more than a body with less mass.

If weight (the pulling force on a body) is proportional to the mass of that body, inserting a body's mass and its weight into the equation $F = ma$ (Newton's second law; see Chapter 3) presents an outcome with important ramifications to mechanics. Substitute W for F in the equation:

$$W = ma$$

Isolate for a:

$$a = W \div m$$

Because W, the weight, is proportional to m, the mass, the value a, the acceleration, will be the same for every body. This acceleration — the acceleration caused by the pull of the Earth on a body — is called the gravitational acceleration, represented with g. And the value of g is assigned 10 m/s/s downward, or –10 m/s/s. (Actually, it's assigned a value of –9.81 m/s/s, but I'll round it off to –10 m/s/s for convenience. What's a little rounding off among friends?)

Slipping, Sliding, and Staying Put: Friction Is FμN

Friction force is present when two bodies contact and want to slide, or are sliding, across each other. Friction always acts to resist sliding; sometimes the friction can be so large that it prevents sliding. It's the "stickiness" between two surfaces. Friction is obvious when you rub your hands together, with the warmth from the friction a good thing on a cool day. When you walk or jump and land, you notice (sometimes painfully) when the available friction is not enough to prevent a slip. But you don't always appreciate the importance of friction during most activities (although either too much or too little friction can adversely affect performance and cause injury).

The four characteristics of a force apply to friction as they do to any force:

- **Magnitude:** The magnitude of friction varies as the objects tend to slip or slide across each other. For every instance of contact between two bodies, there is a maximum friction force that can be developed before the bodies will start to slide across each other.

- **Direction:** The direction of friction is always opposite to the possible direction of sliding between the bodies.

- **Point of application:** The point of application of friction is always within the area of contact between the two bodies.

- **Line of action:** The line of action of the friction force is an extension of the force vector, parallel to the surfaces in contact.

Fill your backpack with books, and set it on a table. Draw an FBD of the backpack sitting at rest on the table (you don't have to draw an actual backpack, unless you're a great sketch artist — a simple square will do the trick). Your drawing should look like Figure 4-10. The two external forces are the weight of the backpack acting downward and the force of the table pushing up on the backpack.

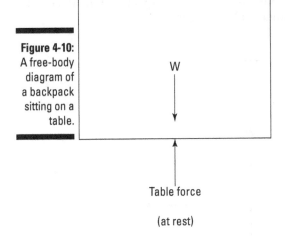

Figure 4-10:
A free-body
diagram of
a backpack
sitting on a
table.

W

Table force

(at rest)

Try to push the backpack sideways to the right, but apply just a light force. Your push is the called the motive force because it is the force tending to cause the backpack (or any body) to slide. Add your pushing force to the FBD; it should look like Figure 4-11. Because your push on the backpack didn't make the backpack slide, there must be a force opposing your push. Where, oh where, could this force be?

Figure 4-11: A free-body diagram of a backpack with a slight push to the right.

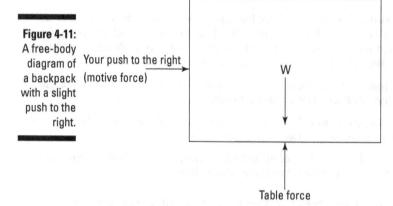

Because no sliding occurred, there must be a force *opposing* the motive force applied by your hand. The size of the opposing force must be the same size as your push to the right because there are no unbalanced forces acting to cause the body to change motion. The direction of the force must be, obviously, opposite to the direction of your pushing force to the right, so the direction of the opposing force must be to the left. This opposing force is friction, and it's applied at the contact between the backpack and the table. Add this force into your FBD; it should look like Figure 4-12.

Figure 4-12: A free-body diagram of a backpack with a slight push to the right, showing the force of friction.

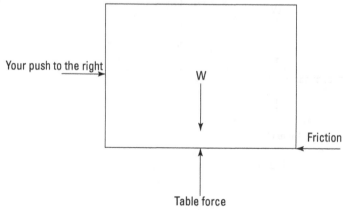

Push a little harder, but still not hard enough to cause the backpack to slide. The opposing force increased as your pushing force increased. Finally, push even harder, and you can get the backpack to slide. Because your pushing force caused the backpack to slide, the friction force must have reached a maximum magnitude above which it could get no greater.

Figure 4-13 graphs what happens in this scenario. As you push harder, the friction force increases to match your push. This relationship is linear, because the magnitude of the motive force and the friction force match exactly. The friction is limiting the sliding, so it's called *limiting friction* (also known as *static friction*). However, at some point, the friction force can increase no more. The peak is called the *maximum limiting friction* (also known as *maximum static friction*). If you push on a body with a force greater than the maximum limiting friction, the body will start to slide.

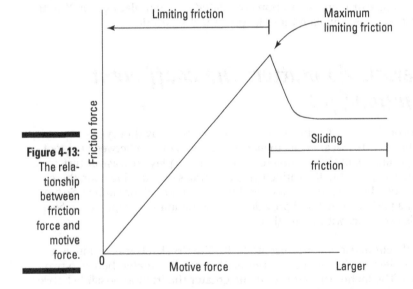

Figure 4-13: The relationship between friction force and motive force.

The amount of friction force actually drops below the peak as the body begins to slide. The friction present when a body slides is called *sliding friction*. You may have sensed this when trying to slide something heavy, like a sofa. You push hard on it in the direction you want the sofa to slide, but friction resists your pushing force. You push harder and harder until the sofa finally starts to slide (you've exceeded the maximum limiting friction). Then you notice that you don't have to push as hard on the sofa to keep it sliding to its new position in the room (because the friction present has decreased).

Friction is the force that resists sliding. In some situations, such as maintaining traction when walking or running, or when holding a tennis racket to hit the ball without your hand slipping on the racket handle, a relatively high value of maximum limiting friction is desired. In other situations, such as when a gymnast performs a swing around the high bar, or when pivoting on the ball of your planted foot, a relatively low value of maximum limiting friction is desired.

The magnitude of the limiting friction force is affected by two factors:

- The materials of the two bodies in contact with each other (μ)
- The force squeezing the two bodies together (N)

Friction is FμN, because the equation for the magnitude of friction is $F = \mu N$. The friction force will increase if μ is greater or if N is greater. Techniques to either decrease or increase friction are related to manipulating μ or N. In the following sections, I explain these factors affecting friction.

Materials do matter: The coefficient of friction (μ)

Do cleats on the sole of a shoe increase friction? You may think yes — everyone knows that athletes wear cleats to increase traction or grip between the foot and the ground, right? When athletes wear cleats to play on grass or turf, they get greater traction. But if an athlete in cleats walks on a marble floor, there is *less* traction between the cleats and the hard, smooth floor surface. So, you can't say a particular material provides a lot of friction — it depends on *both* materials in contact with each other.

The coefficent of friction, represented with the Greek letter μ, or mu, is a number reflecting the amount of friction that can be created between two materials. The higher the value of μ, the greater the friction possible between the two surfaces. Table 4-1 presents example μ values for different pairs of materials.

The high μ values for tires and a clothed body on dry pavement reflect the high friction present in both scenarios. The high μ value for the tire is desirable, to provide grip between the tire and the road. The high μ value for a clothed body on dry pavement reveals why an unrestrained motorist can suffer "road burn" if ejected to land on the road. The low μ value for steel on ice indicates why skaters glide easily on ice, and the extremely low μ value for healthy human cartilage reveals why healthy joints move so smoothly.

Table 4-1   Example μ Values for Different Pairs of Materials

Materials	μ Values
Tire on dry pavement	0.9
Clothed body on dry pavement	0.7
Soccer shoes on artificial grass, going forward	0.8
Soccer shoes on artificial grass, going sideways	0.6
Ballet shoes on linoleum	0.4
Ballet shoes on hardwood	0.3
Steel on ice	0.1
Healthy human cartilage	0.009

The two values presented for soccer shoes on artificial grass reflect how cleats are aligned on the bottom of shoes. Although the material of the shoe is the same, the cleat alignment affects how the shoe interacts with the grass when pushing off or stopping in the forward or backward direction, and when cutting in the sideways direction. One problem with designing cleats to interact with the playing surface: If the friction is too high, the foot may become fixed on the ground and the risk of knee injury is increased.

Squeezing to stick: Normal reaction force (N)

When two lines are at 90 degrees to each other, we say the lines are *perpendicular*. When two surfaces are at 90 degrees to each other, we say the surfaces are *orthogonal*. When a surface and a line are at 90 degrees to each other, we say the line is *normal to the surface*. Because a vector is just an embellished line, when a force is acting at 90 degrees to a surface, we say the force is *normal*.

Consider the backpack at rest in Figure 4-10. The force from the table is pushing upward on the backpack, at 90 degrees to the backpack. Thus, this is a normal reaction force, and we can call this N.

N is the force of interest in determining how much friction can be present. N is the force "squeezing the bodies together," and N affects the magnitude of the friction force produced between the box and the surface.

Increasing N will increase the magnitude of friction that can be developed to resist sliding. Place this book on a table, and push down on this book as you also push sideways on the book. It's harder to slide than it would be if you simply pushed sideways without a downward push. The materials in contact have not changed, but N has been increased in response to the downward push. A tight grip on a tennis racket can prevent the racket from slipping when it contacts the ball, because the tight grip increases the N squeezing the racket and hand together, increasing friction. Conversely, decreasing N will make something slide easier by reducing the friction present. Pushing up on the sofa as you push sideways makes the sofa easier to slide because your upward push decreases N.

μ is actually a proportion between the maximum limiting friction and the normal reaction force. μ is calculated quite simply:

1. **Put a block of known weight on a table.**

2. **Tilt the table.**

3. **Carefully identify the angle of tilt when the block begins to slide.**

    The tangent of that angle is μ.

Or simply put the block of known weight on a flat table, and pull on the block as you measure the pulling force. Record the force at the instant the body begins to slide, and calculate:

μ = Pulling Force ÷ Object Weight

Chapter 5

Describing Linear Motion: Linear Kinematics

. . . .

In This Chapter

▶ Describing position in a reference frame

▶ Defining change of linear position

▶ Computing how fast a body moves

▶ Introducing momentum

▶ Describing acceleration

▶ Using the equations of constant acceleration to solve projectile motion

. . . .

*L*inear kinematics is the branch of mechanics describing linear motion, or *translation*. In linear motion, all points on a body go equally far in the same direction, they travel equally fast, and they move at the same time. All points on the body also speed up and slow down at the same time. There are two forms of linear motion:

▸ **Rectilinear motion:** Motion in a straight line.

▸ **Curvilinear motion:** Motion along a curved line, such as when a body moves through the air as a projectile. Curvilinear motion is simultaneous motion in the up–down (vertical) and forward–backward (horizontal) directions.

In this chapter, I explain the kinematic descriptors to describe where a body is in space (position) and how far, how fast, and how consistently (speeding up or slowing down) the body moves. I introduce the important quantity of linear momentum because it combines the two critical ideas of the body's current state of motion and the body's resistance, or inertia, to changing its motion. Finally, I demonstrate the use of the three equations of constant acceleration to quantify projectile motion.

Identifying Position

Position describes a body's location in space. The location is specified relative to a selected landmark. The landmark is referred to as the *origin,* because all measures of position are made from, or originate from, this point. Figure 5-1 shows a view from above of a player on a soccer field (or pitch, to be sport-term correct) at a specific instant in time. I've set the origin to a coordinate system (see Chapter 2) where the goal line and the sideline meet in the bottom-left corner of the field; the *x* direction corresponds to the length of the field, and the *y* direction corresponds to the width of the field. I'll use this two-dimensional coordinate system to specify the player's position on the field in both one and two dimensions:

- ✔ **One dimension:** 30 m from the goal line (the forward–backward dimension alone, *x* = 30 m) *or* 15 m from the sideline (across the field dimension alone, *y* = 15 m)

- ✔ **Two dimensions:** 30 m from the goal line *and* 15 m from the sideline (the forward–backward dimension *and* the side-to-side dimension specified at the same time, *x* = 30 m *and y* = 15 m, or simply (30,15), using the typical method of reporting the *x* and *y* positions on a coordinate system)

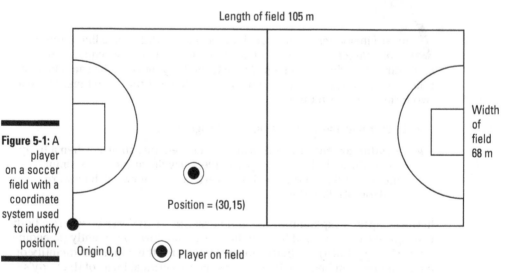

Length of field 105 m

Width of field 68 m

Figure 5-1: A player on a soccer field with a coordinate system used to identify position.

Position = (30,15)

Origin 0, 0   ◉ Player on field

A third dimension, the player's position above the field, could also be described. The coordinate system would be modified to include a third dimension, in the *z* direction. This *z* dimension would be aligned up and down. A three-dimensional description of the player's position would specify the *x, y,* and *z* coordinates of the player. In this book, I stick to two-dimensional descriptions of linear kinematics, although all the descriptors can be extended to three dimensions.

Describing How Far a Body Travels

Setting up a coordinate system with a specified origin allows me to describe a body's position, or location, in space. The next step is describing the change in location that occurs when a body moves in space, a description of how far the body travels. *Distance* and *displacement* are often used interchangeably in everyday conversation, but they have different, and specific, meanings in kinematics when used to describe a change of position.

Figure 5-2 shows the soccer player's position on the field at two instants, 17 seconds (s) and 39 s into the game. The original position (30,15), described in the preceding section, is identified as p_1, for position 1. The player dekes and dodges while running on the field, and gets to p_2, for position 2. At p_2, the player is at (55,45); the player is now 55 m from the goal line and 45 m from the sideline. The path followed by the player is the squiggly line from p_1 to p_2. How far did the player go? The distance and displacement traveled are explained in this section.

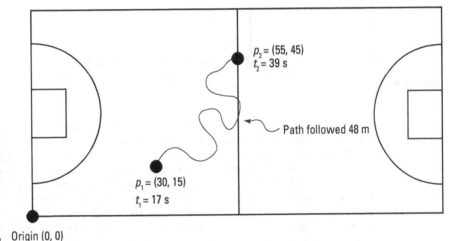

Figure 5-2:
The soccer player has moved from p_1 to p_2 in 22 seconds.

$p_2 = (55, 45)$
$t_2 = 39$ s

Path followed 48 m

$p_1 = (30, 15)$
$t_1 = 17$ s

Origin (0, 0)

Distance

Distance is the length of the path followed during the change of position from p_1 to p_2, without considering the direction of travel. Distance is represented with ℓ, for length of path followed. The path followed is 48 m, as indicated with the squiggly line between p_1 (30,15) and p_2 (55,45); $\ell = 48$ m. Distance is a scalar quantity and is fully described by specifying magnitude (for distance, that's the length of the path followed).

Displacement

Displacement is the length of the straight line traveled in a specified direction between p_1 and p_2. Displacement is a vector quantity, so I use an arrow to represent displacement. The arrow shows both "how far" and "in what direction" the change in position occurred (see Chapter 2 for more on vectors). In Figure 5-3, I zoom in on the section of the soccer field of interest to explain displacement more clearly.

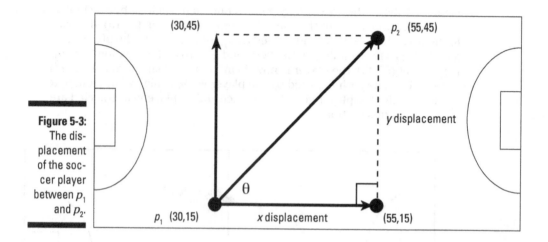

Figure 5-3:
The dis-
placement
of the soc-
cer player
between p_1
and p_2.

In Figure 5-3, three displacements are represented with vectors. The resultant displacement vector directly joins p_1 and p_2. As with any vector, the resultant displacement can be resolved into two components at right angles to each other. For the soccer player, these component vectors represent the change of position along the length of the field (displacement in the *x* direction) and the change in position between the sidelines (displacement in the *y* direction). The displacement vectors in the *x* and *y* directions are shown in Figure 5-3. The change of position along the length of the field (*x* direction) goes from (30,15) to (55,45); the *y*-coordinate is the same, so this vector represents how far the player traveled in the *x* direction (*x* displacement). The change of position between the sidelines (*y* direction) goes from (30,15) to (30,45); the *x*-coordinate is the same, so this vector represents how far the player traveled in the *y* direction (*y* displacement).

In a coordinate pair like (30,15), the first value (30) is the *x*-coordinate and the second value (15) is the *y*-coordinate.

Displacement is represented with $\Delta p_{\text{direction}}$, with the Greek letter Δ (delta) the symbol for "change in" and the subscript word *direction* indicating the direction in which the displacement is calculated. Δp_x is displacement in the *x* direction, and Δp_y is displacement in the *y* direction.

$\Delta p_{direction} = p_f - p_i$ is the equation for displacement, where p_f is the final position in the specified direction (the position at the end of the phase being analyzed) and p_i is the initial position in the specified direction (the position at the start of the phase being analyzed). The units for displacement are the units used to identify the position.

To calculate displacement in the x direction, use these two steps:

1. **Identify the initial and final x coordinates.**

    • $p_i = 30$ m (use the x value from the start position coordinates (30,15))

    • $p_f = 55$ m (use the x value from the final position coordinates (55,45))

2. **Select the equation for displacement and substitute the values.**

    $\Delta p_x = p_f - p_i = 55$ m $- 30$ m $= 25$ m

Use the same two steps to calculate displacement in the y direction:

1. **Identify the initial and final y-coordinates.**

    • $p_i = 15$ m (use the y value from the start position coordinates (30,15))

    • $p_f = 45$ m (use the y value from the final position coordinates (55,45))

2. **Select the equation for displacement and substitute the values.**

    $\Delta p_y = p_f - p_i = 45$ m $- 15$ m $= 30$ m

The displacement of the player is described as +25 m in the x direction and +30 m in the y direction.

Resultant displacement is calculated using the displacements in the x and y directions. In Figure 5-3, I've drawn a dashed line joining the tips of the x vector and the resultant vector and a dashed line joining the tips of the y vector and the resultant vector. The dashed lines are parallel and equal in length to the vectors opposite, creating a parallelogram. More important, because a right triangle is created, I can use the Pythagorean theorem and the trig function arctan (see Chapter 2) to solve for the magnitude and direction of the resultant displacement using these steps:

1. **Select the right triangle to use.**

    The best right triangle to use is the one with the right angle created between the vector of the x displacement and the dashed line equal to the y displacement; the resultant displacement is the hypotenuse of this right triangle. The direction of the resultant displacement is the angle between the hypotenuse and the x vector, so this is labeled θ. Figure 5-4 shows this right triangle and the labels.

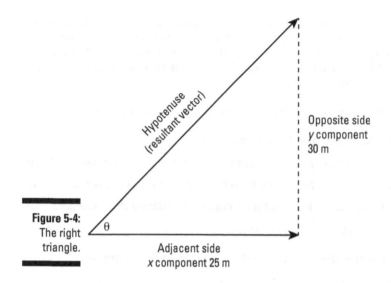

Figure 5-4:
The right
triangle.

Hypotenuse
(resultant vector)

Opposite side
y component
30 m

θ

Adjacent side
x component 25 m

2. **Use the Pythagorean theorem to solve for the magnitude of the resultant displacement.**

$$\text{Resultant}^2 = x \text{ component}^2 + y \text{ component}^2$$
$$\text{Resultant} = \sqrt{x \text{ component}^2 + y \text{ component}^2}$$
$$= \sqrt{625 + 900} = \sqrt{1,525} = 39.1 \text{ m}$$

3. **Solve the direction of the resultant displacement using the arctan function.**

$$\theta = \arctan \frac{y \text{ component}}{x \text{ component}} = \arctan \frac{30}{25} = \arctan 1.5 = 56.3°$$

The resultant displacement is 39.1 m at an angle of 56.3 degrees.

Describing How Fast a Body Travels

Change in position means a body moves in space during a period of time. Distance and displacement describe how far the body moves in space. The time period is called Δt (read "delta t"). Δt is calculated as $t_f - t_i$ (read "time final minus time initial"), where t_f is the time at the end of the period and t_i is the time at the start of the period. Dividing "how far" by "time period" gives a measure of how fast a body moves, the rate of change of position per unit of time. The units of "how fast" are units of "how far" divided by units of "time," or m/s (meters per second). You're probably familiar with mph (miles per hour) as units of how fast you're traveling in a car.

Speed and *velocity* are the terms to describe how fast a body travels. Like *distance* and *displacement,* the terms are often used interchangeably in everyday conversation, but they have different, and specific, meanings in kinematics. In this section, I explain the calculations of speed and velocity.

Speed

Speed is how fast a body moves, or the body's rate of motion. The direction of motion is not considered. Speed is represented with the letter s and is defined as the distance (ℓ) covered during the change in position divided by the time taken to move the distance (Δt). As an equation, $s = \dfrac{\ell}{\Delta t}$. Δt is calculated as $t_f - t_i$ (read "time final minus time initial"), where t_f is the time at the end of the period and t_i is the time at the start of the period. Speed is a scalar quantity and describes only how fast a body moves, or the body's rate of motion. The direction of the change in position is not taken into consideration to calculate speed. (What a relief for the folks preparing to time travel!)

For the player in Figure 5-2, the path traveled is 48 m, the length of the squiggly line between p_1 (30,15) and p_2 (55,45); $\ell = 48$ m. The clock read 17 s when player was at p_1 and 39 s when the player was at p_2. To solve for speed, follow these steps:

1. **Identify the known values and create a table of variables.**

   - $t_f = t_2 = 39$ s
   - $t_i = t_1 = 17$ s
   - $\ell = 48$ m
   - **Speed** = Unknown, to be solved

2. **Select the equation and substitute the given values.**

$$s = \frac{\ell}{\Delta t} = \frac{\ell}{t_f - t} = \frac{\ell}{39\ \text{s} - 17\ \text{s}} = \frac{48\ \text{m}}{22\ \text{s}} = 2.2\ \text{m/s}$$

The player's speed was 2.2 m/s.

Speed is a scalar quantity, so there is no mention of direction, only a description of how fast the player moved.

Velocity

Velocity describes how fast a body moves in a specific direction. To calculate velocity, the displacement of a body in a specific direction ($\Delta p_{\text{direction}}$) is divided by the time it took for the displacement to occur (Δt), or as an equation $v_{\text{direction}} = \frac{\Delta p_{\text{direction}}}{\Delta t}$. Velocity is a vector quantity and describes both how fast a body moves and in what direction the body moves. The direction must be stated to describe velocity.

For the player in Figure 5-2, three displacements were calculated above describing the change in position between p_1 (30,15) and p_2 (55,45): the resultant displacement of 39.1 m at $\theta = 56.3$ degrees, and the component displacements of 25 m in the x direction (down the field) and 30 m in the y direction (across the field). The same steps are used to solve for the velocities in the resultant x and y directions. I'll start with solving for the resultant velocity:

1. **Identify the known values and create a table of variables.**

   - $t_f = t_2 = 39$ s
   - $t_i = t_1 = 17$ s
   - $\Delta p = 39.1$ m
   - $\theta = 56.3$ degrees
   - $v_{\text{resultant}} =$ Unknown, to be solved

2. **Select the equation and substitute the given values.**

$$v_{\text{resultant}} = \frac{\Delta p_{\text{resultant}}}{\Delta t} = \frac{\Delta p_{\text{resultant}}}{t_{fi} - t} = \frac{39.1 \text{ m}}{39 \text{ s} - 17 \text{ s}} = \frac{39.1 \text{ m}}{22 \text{ s}} = 1.8 \text{ m/s}$$

The player's resultant velocity was 1.8 m/s at an angle of 56.3 degrees. Velocity is a vector quantity, so both magnitude (1.8 m/s) and direction (56.3 degrees) must be described.

And now for the velocity in the x, or down-the-field, direction:

1. **Identify the known values and create a table of variables.**

   - $t_f = t_2 = 39$ s
   - $t_i = t_1 = 17$ s
   - $p_i = 30$ m (use the x value from the start position coordinates (30,15))
   - $p_f = 55$ m (use the x value from the final position coordinates (55,45))
   - $v_x =$ Unknown, to be solved

2. Select the equation and substitute the given values:

$$v_x = \frac{\Delta p_x}{\Delta t} = \frac{p_f - p_i}{t_{fi} - t} = \frac{55 \text{ s} - 30 \text{ s}}{39 \text{ s} - 17 \text{ s}} = \frac{25 \text{ m}}{22 \text{ s}} = 1.1 \text{ m/s}$$

The player's velocity was 1.1 m/s in the *x*, or down-the-field, direction.

If $\Delta p_{\text{direction}}$ and/or Δt are already solved, use those values when calculating velocity. I did this with $v_{\text{resultant}}$, but for the *x* and *y* velocities I'm showing all the steps in full.

And finally for the velocity in the *y*, or across-the-field, direction:

1. Identify the known values and create a table of variables.

- $t_f = t_2 = 39$ s
- $t_i = t_1 = 17$ s
- $p_i = 15$ m (use the *y* value from the start position coordinates (30,15))
- $p_f = 45$ m (use the *y* value from the final position coordinates (55,45))
- $v_y =$ Unknown, to be solved

2. Select the equation and substitute the given values.

$$v_y = \frac{\Delta p_y}{\Delta t} = \frac{p_{fi} - p}{t_{fi} - t} = \frac{45 \text{ s} - 15 \text{ s}}{39 \text{ s} - 17 \text{ s}} = \frac{30 \text{ m}}{22 \text{ s}} = 1.4 \text{ m/s}$$

The player's velocity was 1.4 m/s in the *y*, or across-the-field, direction.

Describe the velocity of the player like this: The resultant velocity of 1.8 m/s at 56.3 degrees consisted of component velocities of 1.1 m/s in the *x* direction and 1.4 m/s in the *y* direction.

The speed or velocity calculated with the equations is the *average* speed or the *average* velocity of the body over the entire period Δt. The speed and velocity may actually vary over that interval. To calculate the instantaneous speed or velocity, or how fast the body is moving at an instant in time, the time period Δt is made very small. In a quantitative analysis of performance (see Chapter 17), when technology is used to record and analyze performance, a Δt of 0.005 seconds is commonly used in the calculations.

Momentum

Momentum (M) is the product of mass and velocity; as an equation, that's $M = mv$. The units are kg · m/s (read "kilogram meters per second"). Mass is the measure of a body's *inertia,* or resistance to changing its current state of motion. Velocity describes the current state of motion of a body, how fast it's moving in a specific direction. As a single quantity, momentum describes the current state of motion of a body and its resistance to changing that motion.

Consider a player with 100 kg mass moving at 4 m/s. To calculate the momentum of the player, follow these steps:

1. **Identify the known values and create a table of variables.**

   • m = 100 kg

   • Velocity = 4 m/s

   • Momentum = Unknown, to be solved

2. **Select the equation and substitute the given values.**

   $M = mv = 100$ kg · 4 m/s $= 400$ kg · m/s

   If the velocity is 4.5 m/s, $M = 100$ kg · 4.5 m/s $= 450$ kg · m/s.

   If the velocity is 3.5 m/s, $M = 100$ kg · 3.5 m/s $= 350$ kg · m/s.

When velocity increases, the momentum of a body increases. When velocity decreases, the momentum of a body decreases. In most human performance, the mass of the body stays the same during the performance, and changes in momentum reflect changes in the velocity of the body. In Chapter 6, I explain the benefit of looking at all performance as the change in momentum of the body under analysis.

Speeding Up or Slowing Down: Acceleration

Acceleration (a) describes the rate of change of velocity with respect to time. To calculate acceleration, the change in the velocity ($\Delta v_{direction}$) of a body during a period of time is divided by the length of time (Δt), or as an equation $a = \frac{\Delta v}{\Delta t}$. Acceleration quantifies the concept of "change in motion," and a body accelerates when it speeds up or slows down in a particular direction.

Δv, the change in velocity during the period being studied, is calculated as $v_f - v_i$. Δt is the duration of the period. Both velocities used to calculate Δv must be in the same direction — I could calculate the difference between velocity in the x direction and velocity in the y direction, but this would be not only wrong but also meaningless.

TIP

The units of acceleration are the units of velocity (m/s) divided by units of time (seconds), or meters per second per second (m/s/s).

Figure 5-5 shows a person jumping up and coming back down. The person is shown at five time instants during this task:

- ✔ At t_1, the person is crouched and at rest ($v_{\text{vertical}} = 0$).

- ✔ At t_2, the person has just left the ground with $v_{\text{vertical}} = +2$ m/s.

- ✔ At t_3, the person has stopped moving upward and has not yet started to move downward, so $v_{\text{vertical}} = 0$.

- ✔ At t_4, the person has just touched the ground with $v_{\text{vertical}} = -2$ m/s.

- ✔ At t_5, the person has come to rest in a crouched position, with $v_{\text{vertical}} = 0$ m/s.

Figure 5-5: A person jumps straight up and then down.

	Take-off	Peak height	Touch down	
$t_1 = 0$	$t_2 = .5$	$t_3 = .7$	$t_4 = .9$	$t_5 = 1.5$
$v_1 = 0$	$v_2 = +2$	$v_3 = 0$	$v_4 = -2$	$v_5 = 0$

I'll calculate acceleration over two phases as the person goes up (the ascent phase): as v increases (speeds up) from 0 to +2 m/s, and then as v decreases (slows down) from +2 m/s to 0. Similarly, I'll calculate acceleration over two phases as the person comes down (the descent phase): as v increases (speeds up) from 0 to –2 m/s, and then as v decreases (slows down) from –2 m/s to 0. These calculations are shown in the following table:

Interval	v_i	t_i	v_f	t_f	$\Delta v = v_f - v_i$	$\Delta t = t_f - t_i$	$a = \dfrac{\Delta v}{\Delta t}$
Ascent							
t_1 to t_2	0	0	+2	0.5	$2.0 - 0 = +2.0$	$0.5 - 0 = 0.5$	$+2.0 \div 0.5 = +4.0$
t_2 to t_3	+2	0.5	0	0.7	$0 - (+2.0) = -2.0$	$0.7 - 0.5 = 0.2$	$-2.0 \div 0.2 = -10.0$
Descent							
t_3 to t_4	0	0.7	–2.0	0.9	$-2.0 - 0 = -2.0$	$0.9 - 0.7 = 0.2$	$-2.0 \div 0.2 = -10.0$
t_4 to t_5	–2.0	0.9	0	1.5	$0 - (-2) = +2$	$1.5 - 0.9 = 0.6$	$+2.0 \div 0.6 = +3.3$
Units	m/s	s	m/s	s	m/s	s	m/s/s

Both the ascent and descent phases show periods of + and – acceleration. But the interpretation of the + and – accelerations is opposite in each phase. When rising up, moving in the + direction, +a reflects speeding up from 0 to 2 m/s, and –a reflects slowing down from +2 to 0 m/s. When descending, moving in the – direction, –a reflects speeding up from 0 to –2 m/s, and +a reflects slowing down from –2 to 0 m/s.

Unlike Δp and v, the sign of the calculated acceleration does not immediately give information on the motion. The direction of the velocity must also be considered. Three rules for interpreting acceleration are

✔ When the direction of travel and the sign of a are the same, the v increases (the body is speeding up).

✔ When the direction of travel and the sign of a are opposite, the v decreases (the body is slowing down).

✔ Whenever $a = 0$, regardless of the direction of travel, the v stays the same (the body is in a constant state of motion).

In mechanics, the term *acceleration* means a change in the motion, or velocity, of the body. This can be either speeding up or slowing down. So acceleration refers to speeding up or slowing down the velocity of a body. Use the sign of the acceleration and the direction of the motion to interpret a calculated acceleration to determine if a body is speeding up or slowing down.

Constant acceleration

Constant acceleration means acceleration has the same value over a period of time. When acceleration is constant, three equations relate the kinematic quantities of position, displacement, velocity, acceleration, and time. For working with the equations of constant acceleration, I use the following kinematic symbols:

- ✔ u = initial velocity
- ✔ v = final velocity
- ✔ p = displacement
- ✔ a = acceleration
- ✔ t = time, short for Δt

When working with constant acceleration, I always create a table listing the symbols in the order u, v, p, a, and t.

The equations of constant acceleration are

$$v = u + at$$

$$v^2 = u^2 + 2ap$$

$$p = ut + \frac{1}{2}at^2$$

Each equation relates three of the four quantities of displacement, velocity, acceleration, and time. Creating a table of the kinematic symbols, and filling in the values for the known quantities, helps me to select the best equation to solve for the unknown quantity. The equations are most useful for working with projectile motion, as explained in the next section.

Projectile motion

What do a basketball, a long jumper, and a diver have in common? They all become projectiles. A projectile is a body that has been projected or dropped into the air. To be a projectile, the body cannot be in contact with another body. Gravity (see Chapter 4) acts in the vertical direction and causes an acceleration of –10 m/s/s (downward). If we ignore air resistance (see Chapter 9), the acceleration in the horizontal direction is 0 m/s/s.

The value of g, or gravitational acceleration, is actually –9.81 m/s/s. But I use –10 m/s/s for ease of calculations. (Gravitational acceleration from the weight of a body is explained in Chapter 4.)

Figure 5-6 shows a ball kicked into the air. The ball is a projectile from the instant it leaves the kicker's foot until it hits the ground again. A parabola is an inverted U, and this shape of flight path is created because the projectile moves simultaneously in the horizontal and vertical directions. Figure 5-6 shows some important descriptors of the ball's parabolic path while it's a projectile:

- v_{TO}: The velocity at takeoff, or how fast and in what direction (θ) the ball is moving at the instant it becomes a projectile. The v_{TO} is the resultant velocity of the ball, and the components of v_v for the vertical velocity and v_h for the horizontal velocity can be resolved from v_{TO} using the trig functions (see Chapter 2).

- $h_{release}$: The height of release, or how high the ball is above the ground when it becomes a projectile.

- h_{flight}: The vertical displacement of the ball from release until it stops moving upward, when $v_v = 0$.

- h_{peak}: The height of the projectile at the peak of the parabolic path it follows. If a body is dropped, h_{peak} is the height from which the body was dropped.

Peak height is the height above the ground. This height is the sum of how high the body moves upward after release (h_{flight}) and $h_{release}$, the height of the body above the ground when it became a projectile.

- v_{vpeak}: The vertical velocity at h_{peak}. Since the projectile has stopped moving up and has not yet started to fall back down, v_{vpeak} is *always* 0.

- $h_{contact}$: The height of contact, or how high the body is above the ground when it ends motion as a projectile. If the projectile lands on the ground, $h_{contact} = 0$.

- t_{air}: The time in the air. The longest time in the air is from the instant of takeoff until the body contacts another body. t_{air} is the sum of t_{up}, from instant of takeoff to peak height, and t_{down}, from peak height until the end of the projectile motion.

- Range: The range is the horizontal displacement of the projectile while it's in the air.

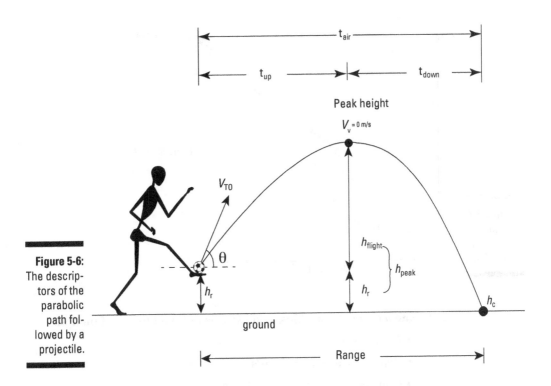

Figure 5-6:
The descriptors of the parabolic path followed by a projectile.

I can use the equations of constant acceleration to describe the parabolic, or inverted U, path of a body while it's a projectile. When working with projectile motion, I use a set series of steps to keep the process orderly. To show these steps, I'll set v_{TO} to be 26 m/s at an angle of 55 degrees and $h_{release}$ to 0.7 meters. I'll use the steps to solve for some vertical motion descriptors (how high the ball goes, how much time it takes to reach peak height, how long it takes to fall to the ground) and for the range, or horizontal displacement, of the ball.

First, resolve the v_{TO} of 26 m/s at an angle of 55 degrees into v_v and v_H, using the trig functions (Chapter 2):

1. **Start by sketching v_{TO} and its components v_V and v_H (see Figure 5-7).**

   I usually solve for v_V first. Note that v_V is the opposite side of the right triangle.

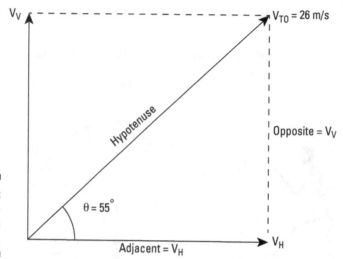

Figure 5-7:
The components v_V and v_H of v_{TO}.

2. **Create a table of variables, and use this to select the trig function.**

   - $\theta = 55$ degrees

   - Hypotenuse $= v_{TO} = 26$ m/s

   - Opposite $= v_V =$ Unknown, to be solved

3. **Using SOH CAH TOA, select** $\sin\theta = \dfrac{\text{opposite}}{\text{hypotenuse}}$, **isolate for the unknown, or opposite side, fill in the known values, and solve.**

   $$\text{opposite} = \text{hypotenuse} \times \sin\theta = 26 \text{ m/s} \times \sin 55° = 26 \text{ m/s} \times 0.819$$
   $$= 21.3 \text{ m/s}$$

   When the ball leaves the player's foot, it's moving at 21.3 m/s in the vertical direction ($v_V = 21.3$ m/s).

   Now solve for v_H. Note that v_H is the adjacent side of the right angle triangle.

4. **Create a table of variables then select the trig function.**

   - $\theta = 55$ degrees

   - Hypotenuse $= v_{TO} = 26$ m/s

   - Adjacent $= v_H =$ Unknown, to be solved

5. **Using SOH CAH TOA, select** $\cos\theta = \dfrac{\text{adjacent}}{\text{hypotenuse}}$, **isolate for the unknown, or adjacent side, fill in known values, and solve.**

   $$\text{adjacent} = \text{hypotenuse} \times \cos\theta = 26 \text{ m/s} \times \cos 55° = 26 \text{ m/s} \times 0.571$$
   $$= 14.9 \text{ m/s}$$

   When the ball leaves the player's foot, it is moving at 14.9 m/s in the horizontal direction ($v_H = 14.9$m/s).

Sometimes the vertical and horizontal velocities are given in the description of projectile motion. In this case, you can skip the preceding steps and move right on to the following.

Next, sketch a parabola representing the projectile motion, filling in what you know and what you're asked to solve for, as shown in Figure 5-8. The known values are given in the problem, are solved for earlier, or are known about projectile motion (that is, $v_V = 0$ m/s at peak height).

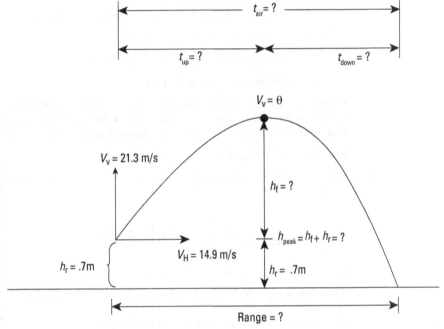

Figure 5-8:
A sketch of the ball's parabolic flight with known and needed values indicated.

A projectile moves simultaneously in the vertical and horizontal directions. Solving first for the descriptors of the vertical motion and then for the descriptors of the horizontal motion, as shown in the next sections, is usually best.

Vertical motion of a projectile

Ignoring air resistance, in the vertical direction the constant acceleration from gravity of –10 m/s/s represents constant acceleration of the projectile. On the way up, the body is slowed down by the acceleration (the sign of the velocity and the sign of the acceleration are opposite). On the way down, the body is sped up (the sign of the velocity and the sign of the acceleration are opposite). Descriptors of the vertical motion of the projectile include how high it goes on the way up (h_{flight}), how long it takes to reach peak height (t_{up}), how long it takes to fall (t_{down}), and the time in the air (t_{air}, the sum of t_{up} and t_{down}).

I'll solve for how high the ball goes (h_{flight}) and use this to solve for h_{peak}. Then I'll solve for t_{up} and t_{down} and use these to calculate t_{air} as $t_{\text{up}} + t_{\text{down}}$. In each solution, I'll show how to apply the equations of constant acceleration, starting with h_{flight}.

1. **Identify the descriptor you're asked to solve for, h_{flight}.**

2. **Create the table of variables (u, v, p, a, and t), filling in known and unknown values.**

   - $u = v_V = 21.3$ m/s (calculated earlier)

   - $v = v_{\text{Vpeak}} = 0$ m/s (vertical velocity is *always* 0 at peak height)

   - $p = h_{\text{flight}} =$ Unknown, to be solved

   - $a = -10$ m/s/s (gravitational acceleration)

   - $t =$ Unknown, but not asked for

3. **Select the equation of constant acceleration linking the three known values and the one unknown value listed in the table of variables, isolate for the unknown, fill in the known values, and solve.**

   Select $v^2 = u^2 + 2ap$ (because t is unknown).

   Set v^2 equal to 0, isolate for p, fill in the known values, and solve the equation:

   $$p = \frac{-u^2}{2a} = \frac{-\left(21.3^2\right)}{2 \times (-10)} = \frac{-453.7}{-20} = 22.7\,\text{m}$$

   The ball goes 22.7 m above the release point.

4. **Solve for $h_{\text{peak}} = h_r + h_{\text{flight}}$, or $h_{\text{peak}} = 0.7 + 22.7 = 23.4$ m.**

   The ball is 23.4 m above the ground at its peak.

Use the steps to solve for the time to reach peak height. To solve for t_{up}:

1. **Create the table of variables (u, v, p, a, and t), filling in known and unknown values.**

   - $u = v_V = 21.3$ m/s

   - $v = v_{\text{Vpeak}} = 0$ m/s

   - $p =$ Unknown, but not asked for

   - $a = -10$ m/s/s (gravitational acceleration)

   - $t = t_{\text{up}} =$ Unknown, to be solved

2. **Select the equation of constant acceleration linking the three known values and the one unknown value listed in the table of variables, isolate for the unknown, fill in the known values, and solve.**

Select $v = u + at$ (because p is unknown).

Set v^2 equal to 0, isolate for t, fill in the known values, and solve the equation:

$$t = \frac{-u}{a} = \frac{-21.3\,\text{m/s}}{-10\,\text{m/s/s}} = 2.1\,\text{s}$$

The ball takes 2.1 s to reach peak height.

Use the steps to solve for the time to fall to the ground. To solve for t_{down}:

1. **Create the table of variables (u, v, p, a, and t), filling in known and unknown values.**

   - $u = V_{\text{vpeak}} = 0$ m/s (the ball starts to fall after it reaches peak height)
   - $v = $ Unknown, and not asked for
   - $p = h_{\text{peak}} = -23.4$ m (the ball falls, so displacement is in the – direction)
   - $a = -10$ m/s/s (gravitational acceleration)
   - $t = t_{\text{down}} = $ Unknown, to be solved

2. **Select the equation of constant acceleration linking the three known values and the one unknown value listed in the table of variables, isolate for the unknown, fill in the known values, and solve.**

Select $p = ut + \frac{1}{2}at^2$ (because v is unknown).

Set u equal to 0, isolate for t, fill in the known values, and solve the equation:

$$t = \sqrt{\frac{2p}{a}} = \sqrt{\frac{2(-23.4\ \text{m})}{-10\ \text{m/s/s}}} = \sqrt{\frac{-46.8\ \text{m}}{-10\ \text{m/s/s}}} = \sqrt{4.68\ \text{s}^2} = 2.2\ \text{s}$$

The ball takes 2.2 s to fall from peak height to the ground.

Since $t_{\text{air}} = t_{\text{up}} + t_{\text{down}}$, substitute in the calculated values to solve: $t_{\text{air}} = 2.1$ s + 2.2 s = 4.3 s. Use this value to solve for the horizontal motion of the ball in the next section.

Horizontal motion of a projectile

In the horizontal direction, ignoring the effect of air resistance, the constant acceleration is 0 m/s/s — there is no change in the horizontal motion of the body. The constant acceleration of 0 m/s/s allows use of the equations of constant acceleration to solve for descriptors of the horizontal motion. The same steps that are used in solving for vertical descriptors are used to solve for horizontal descriptors. I'll show an example by solving for the range, or horizontal displacement, of the ball from release to ground contact.

To solve for range, or horizontal displacement:

1. **Create the table of variables (u, v, p, a, and t), filling in known and unknown values.**

   • $u = V_H = 14.9$m/s (the horizontal component of V_{TO})

   • $v =$ Unknown, and not asked for

   • $p =$ range = Unknown, to be solved

   • $a = 0$ m/s/s (ignoring air resistance)

   • $t = t_{air} = 4.3$ s (as solved for earlier with the vertical descriptors)

2. **Select the equation of constant acceleration linking the three known values and the one unknown value listed in the table of variables, isolate for the unknown, fill in the known values, and solve.**

   Select $p = ut + \frac{1}{2}at^2$ (because v is unknown).

   Set a equal to 0, fill in the known values, and solve the equation:

   $$p = (14.9\,\text{m/s})(4.3\,\text{s}) + \frac{1}{2}(0)(4.3\,\text{s})^2 = 64.1\text{m}$$

The ball has a range, or horizontal displacement, of 64.1 m.

Chapter 6

Causing Linear Motion: Linear Kinetics

• •

In This Chapter

▶ Clarifying net force

▶ Setting Newton's laws of motion as the basis for biomechanics

▶ Deriving the impulse–momentum relationship from Newton's second law

▶ Using the impulse–momentum relationship to calculate the average force on a body

• •

A little over 300 years ago, Sir Isaac Newton published his three laws of
motion. The laws changed the way we look at motion in the universe
because they clearly lay out the cause–effect relationship between the
moving bodies we see and the forces that cause the motion. The laws are the
basis of biomechanics, and this chapter focuses on explaining Newton's laws
and their application to understanding human movement.

In this chapter, I begin by clarifying the ideas of net force and unbalanced
force. Then I present Newton's laws, breaking each law into its fundamental
ideas. Finally, I show how the impulse–momentum relationship provides the
basis for evaluating performance.

Clarifying Net Force and Unbalanced Force

Set a small object such as a book or water bottle on a table. Let's say your
intention is to move the object by applying a force. You can do this a number of
ways. One way is to push the object away from you using your hand. Another
way is to pull the object toward you using your hand. Therefore, we can define
force as a push or a pull acting on a body. A force is an external force when it is

applied from outside of the body under analysis (see Chapter 4). When two or more external forces act in the same direction, the net force is determined by adding the forces together. As an equation, Net Force = $\Sigma F_{direction}$, where the Greek letter Σ (sigma) means to sum the forces, and direction refers to vertical or horizontal. When the forces in a specific direction don't sum to zero, an unbalanced force acts on the body.

Only add external forces together when they act in the same direction. A vertical force of –500 Newtons (N) cannot simply be added to a horizontal force of +100 N because the forces are not applied in the same direction (as indicated by the negative and positive signs). However, the resultant force created by the –500 N vertical force and the +100 N horizontal force can be calculated using the Pythagorean theorem and the arctan trig function (see Chapter 4).

Figure 6-1 shows a free-body diagram (FBD; see Chapter 4) of an individual in contact with the ground at an instant in time. The individual has a mass of 50 kg. The FBD shows four external forces acting on the body:

- **Weight:** 500 N vertical downward

- **Vertical ground reaction force (*GRF$_V$*):** 750 N vertical upward

- **A push to the right:** 200 N horizontal to the right

- **Friction from the ground:** 100 N horizontal to the left

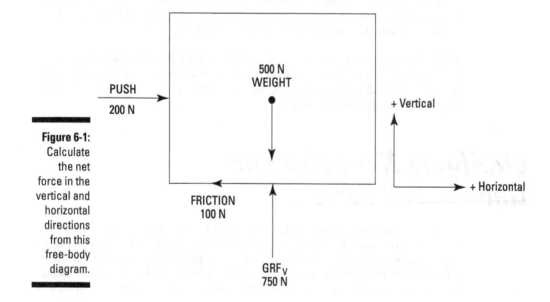

Figure 6-1: Calculate the net force in the vertical and horizontal directions from this free-body diagram.

To calculate the net external force in the vertical and horizontal directions, use the following steps:

1. **Assign a reference system for the direction of the forces.**

   In Figure 6-1, the vertical positive direction is upward, and the horizontal positive direction is to the right.

2. **Label each of the forces.**

   You can assign any name, but it's good practice to name the forces for what they are: weight, GRF_v, push, and friction.

3. **Assign the magnitude (size) and direction (+ or –) to each force.**

   Push = +200 N horizontal, GRF_v = +750 N vertical, friction = –100 N horizontal, and weight = –500 N vertical.

4. **Solve for the net force in the vertical and horizontal directions using $\Sigma F_{\text{direction}}$.**

   The steps are to write out the equation, insert the magnitude and direction of each force in the equation, and then add up the forces in each direction:

Vertical Direction	**_Horizontal Direction_**
$\Sigma F_{\text{Vertical}} = GRF_V + \text{Weight}$	$\Sigma F_{\text{Horizontal}} = \text{Push} + \text{Friction}$
$\Sigma F_V = (+750 \text{ N}) + (-500 \text{ N})$	$\Sigma F_H = (+200 \text{ N}) + (-100 \text{ N})$
$\Sigma F_V = 750 \text{ N} - 500 \text{ N} = +250 \text{ N}$	$\Sigma F_H = 200 \text{ N} - 100 \text{ N} = +100 \text{ N}$

   The net force in the vertical direction is +250 N, or 250 N of force upward. The net force in the horizontal direction is +100 N, or 100 N to the right. Unbalanced forces are acting in both the vertical and the horizontal direction.

When entering the forces into the equation, I always put the magnitude and direction of the force inside brackets, and then I sum the forces. This avoids confusion when summing the forces (trying to remember if a particular force is positive or negative), corresponds with the vector description of a force as having both magnitude and direction, and matches with the idea of "summing the forces."

These steps will always provide the magnitude and direction of the net force in each direction, revealing if an unbalanced force acts on the body. This is one of the most important ideas in mechanics: It's an unbalanced force, and not simply a force, that changes the motion of a body. In the next sections, I explain what Sir Isaac said about unbalanced forces acting on a body in his three laws of motion.

Newton's First Law: The Law of Inertia

Newton's first law specifies unbalanced force as the cause of a change in motion of a body. Because inertia means a body is resistant to changing motion (see Chapter 3), Newton I is also known as the Law of Inertia. Here's what Newton's first law says:

> A body in motion stays in motion and a body at rest stays at rest unless acted upon by an unbalanced external force.

Motion refers to the velocity of a body — how fast a body is moving in a specific direction. (Velocity is explained in Chapter 5.) A body at rest has a velocity of 0. Using velocity in the statement of the law is useful because it shows the implications of Newton's first law: The velocity of a body remains constant if no unbalanced force acts on it. This means a body does not speed up or slow down unless an unbalanced force acts on it. Because acceleration means a change in velocity (see Chapter 5), a body does not speed up or slow down unless an unbalanced force acts on it.

But the law can also be stated to emphasize the cause–effect aspect of the law: Unless an unbalanced force acts on a body, the body does not speed up or slow down. The unbalanced force *causes* the body to accelerate.

It's best to think of motion in each of the three planes or directions: up and down (vertical direction), forward and backward (anteroposterior direction), and side to side (mediolateral direction). A body can move in just one direction (a ball thrown straight up), two directions (a ball rolling at an angle across the surface of a table; it moves forward and to the side at the same time), or simultaneously in all three directions (a ball thrown in the air across a table at an angle; it moves forward, up and down, and to the side, all at the same time). Seeing the motion in the individual directions as contributors to the complete motion is a big step to recognizing how Newton's first law applies. A body can't change its motion in a specific direction unless an unbalanced force acts on the body in that direction.

If the body is speeding up in a specific direction, there must be an unbalanced force pushing or pulling on the body in that direction. On the other hand, if the body is slowing down in a particular direction, there must be an unbalanced force pushing or pulling on the body opposite to the direction it's traveling.

In essence, Newton's first law is a statement of cause and effect: An unbalanced force *causes* a body to speed up or slow down.

I'll skip for a moment Newton's second law, not because it isn't important, but because Newton's third law is a better second step when explaining the laws. (Excuse me for being so bold, Sir Isaac.)

Newton's Third Law: The Law of Equal and Opposite Action–Reaction

People use *action–reaction* in conversations related to politics ("If this law is passed, the public will react this way") and personal relationships ("What goes around comes around"). Newton's third law is often called the *law of action–reaction*, but the repetition of Newton's third law as action–reaction is misleading, because it sounds like one force (an action force) causes the other (a reaction force). The reality is, the two forces occur simultaneously. Each body applies a force on the other body involved in the push or pull. There can't be a force without the two bodies. One body does not cause the force, and the other body does not react to it. The force is there because the two bodies interact. Instead of action–reaction, it's better to think in terms of equal and opposite. A good statement of Newton's third law is:

> When two bodies interact to produce a force, the force on one body is equal in magnitude and opposite in direction to the force on the other body.

Stating Newton's third law this way clarifies the following:

- **There are two bodies involved when force is produced.** Always. This fits with the definition of a force. When one body pulls on another body, the second body pulls on the first body. When you push on the ground, the ground pushes on you. Generally, only one of the bodies is our focus of attention, but a force is applied to both bodies. When an FBD is drawn, the body of interest is separated from the environment and drawn by itself, indicating all the external forces acting on the body.

- **The force on each body is the same size, but the forces are applied in opposite directions on the two bodies. The forces are equal but opposite, creating a mirrored pair.** As you push down on the ground, the ground pushes up on you. When a bat and ball collide, the bat pushes on the ball and the ball pushes on the bat. The mirrored pair of forces acts simultaneously on the two bodies. Figure 6-2 shows a batter hitting a ball, a runner's foot during ground contact, and a person walking into a wall. Below each of these, the equal and opposite forces are shown on each body.

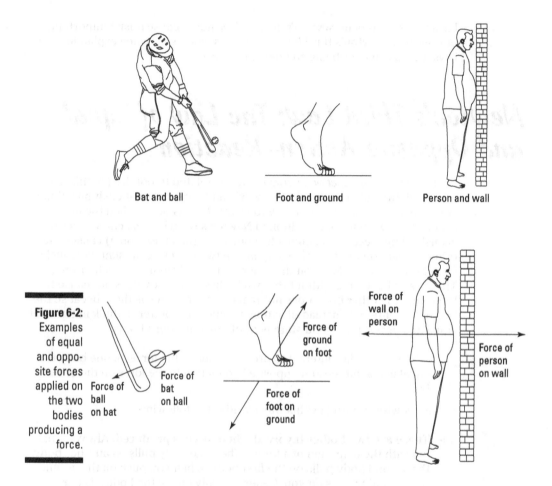

Bat and ball   Foot and ground   Person and wall

Figure 6-2:
Examples
of equal
and oppo-
site forces
applied on
the two
bodies
producing a
force.

Force of
ball
on bat

Force of
bat
on ball

Force of
ground
on foot

Force of
foot on
ground

Force of
wall on
person

Force of
person
on wall

✔ **The law is true even when the bodies are of significantly different mass.** This causes some confusion, because the effect of the equal but opposite forces can't always be seen equally on the two bodies. If you blindly walk into a wall, it's easy to see (and feel) that the wall pushed on you. The force stops you from moving forward and may even cause you to start moving backward. But the wall is still standing there, mind-ing its own business, apparently unaffected. But what if you walk into a padded wall? Again, the force from the wall stops your forward motion — that's easy to feel. But the force you apply to the padded wall deforms the padding in the direction you pushed it.

Seeing that Newton's third law applies to all forces is a matter of mea-surement. If you walk into a solid wall, to the naked eye, the wall is not visibly affected or deformed. But if an instrument sensitive enough to measure small deformations is mounted on the wall, the instrument can detect the deformation caused by the force. Just because we don't see an overt effect of a force doesn't mean that the force isn't there. Newton's third law says the force is there, and who are we to argue with Newton?

Try this, but very carefully (especially the second part): Stand on the floor. Now jump upward and forward. Your motion changed. Because your motion changed, according to Newton's first law, an unbalanced force acted on you. Because you accelerated upward and forward, unbalanced forces acted on you in the vertical and the anteroposterior directions. Easy enough, you pushed on the floor and the floor pushed on you, causing your motion to change. But why didn't the floor move if force was applied to it? This raises doubt about Newton's third law: You moved, but the floor didn't. Now stand at the edge of the seat of a steady chair. Jump upward and forward. The chair pushed on you, and you changed motion. But compared to when you pushed on the floor and the floor didn't move, the chair _did_ move, backward, in the direction you pushed it. The forces were equal in magnitude (your force on the chair was the same size as the force of the chair on you) and opposite (you pushed backward on the chair, the chair pushed forward on you). Newton's third law is not always evident, but it is always in effect.

The chair pushed upward and forward on you, and you accelerated in those directions. But the chair only moved backward, not downward. Why no vertical motion for the chair? Because the floor was pushing up on the chair, so the chair had no unbalanced force in the vertical direction. Without an unbalanced force in the vertical direction, there is no change in the motion of the chair. Your body had an unbalanced force on it in the vertical direction, but the chair didn't. This emphasizes the importance of defining the body of analysis and looking at all the forces acting on it in every direction.

And now onto Newton's second law.

Newton's Second Law: The Law of Acceleration

Newton's second law is the Big Kahuna of mechanics. The law tells what an unbalanced force will do to the motion of a body:

> An _unbalanced_ force applied to a body causes an acceleration of the body with a magnitude proportional to the unbalanced force, in the direction of the unbalanced force, and inversely proportional to the body's mass.

Mathematically, the law is usually thought of as $F = ma$, but this formula is an oversimplification of the law. Newton's second law can be digested in parts, leading to a better mathematical statement of the law, like this:

- ✔ **An *unbalanced* force applied to a body causes an acceleration of the body . . . :** Nothing new here — this restates Newton's first law. In mechanics, *acceleration* means "to speed up or slow down in a specific direction," a change in the motion of the body.

- ✔ **. . . with a magnitude proportional to the unbalanced force . . . :** A larger unbalanced force causes a larger change of motion, and a smaller unbalanced force causes a smaller change of motion.

- ✔ **. . . in the direction of the unbalanced force . . . :** If the unbalanced force is in the positive direction, the acceleration will be in the positive direction. If the unbalanced force is in the negative direction, the acceleration will be in the negative direction. The direction of the unbalanced force and the direction of the acceleration caused by the unbalanced force are the same.

TIP

In Chapter 5, I explain how the sign of the velocity and the sign of the acceleration must be used to interpret the effect of a calculated acceleration. Newton's second law states that the direction of the acceleration is the same as the direction of the unbalanced force. Combining this with the rules to interpret acceleration shows that if the unbalanced force acts in the same direction that the body is already moving, the body will speed up (the sign of the velocity and sign of the acceleration are the same). If the unbalanced force is *opposite* to the direction that the body is already moving, the body will slow down (the sign of the velocity and the sign of the acceleration are opposite). If the body is at rest, the body will start to move in the same direction as the unbalanced force.

Beautiful. This confirms what you probably already know. To make a ball move faster, kick it harder (apply a larger force). To stop a ball, put your foot in front of the ball and push back opposite to its direction of travel. The law also gives insight to why the ball may have gone in the wrong direction from where you wanted: The force you applied to the ball was in the wrong direction. The change of motion of the ball cannot violate Newton's second law.

- ✔ **. . . and inversely proportional to the body's mass:** The size of the acceleration, or change in motion, depends on the mass of the body. An unbalanced force of a given magnitude causes smaller acceleration of a more massive body than it causes on a less massive body. Think about kicking a soccer ball compared to kicking a bowling ball. The bowling ball has greater mass than the soccer ball. If each ball is kicked with the same amount of force, the bowling ball will not go as fast as the soccer ball; the soccer ball accelerates more than the bowling ball. To get the bowling ball and the soccer ball to go at the same speed, you must kick the bowling ball much harder. Note also that kicking a bowling ball fast will confirm Newton's third law. When you kick a soccer ball fast, the force required doesn't cause much discomfort to your foot (but it will if you kick the ball barefoot). Kicking the bowling ball hard enough to make it go fast will require a force large enough to, putting it mildly, cause some discomfort to your foot, whether you're wearing shoes or not.

As I mention earlier, Newton's second law is often expressed simply as $F = ma$ (force equals mass times acceleration). But F must be thought of as the net force, or the sum of the forces (ΣF, where the Greek letter Σ represents "sum of"). And because the direction of the force and the direction of the acceleration are linked, direction needs to be included. Until these important aspects of Newton's second law are fixed in your mind, consider the law as $\Sigma F_{\text{Direction}} = ma_{\text{Direction}}$.

TIP

A great way to write Newton's second law is

$$a_{\text{Direction}} = \frac{\Sigma F_{\text{Direction}}}{m}$$

Written this way, the equation can be read: The acceleration of the body is directly proportional to the magnitude of the unbalanced force, the acceleration of the body is in the direction of the unbalanced force, and the acceleration of the body is inversely proportional to the mass of the body.

Let's rethink the jumping task presented in the preceding section, in light of Newton's second law. When you're on the ground, the force you apply to the ground is equal in magnitude and opposite in direction to the force the ground applies to you. Your mass is much less than the mass of the earth. The force of the ground on you causes the observable change in your motion. But the equally large force applied to the earth by you has no visible effect on the motion of the earth because the earth is *so* massive. When you jumped off the chair, your force on the chair did cause an observable change in its motion because the mass of the chair is much less than the mass of the earth.

The following examples show how to use Newton's second law to calculate the acceleration of a body when the forces acting in a specific direction have been identified. The mass of the body is the same, and F_1 is the same, in each example. I use different magnitude and/or signs for F_2 to show how the net force created by the two forces affects the acceleration:

Example 1	*Example 2*	*Example 3*
$F_1 = +20$ N	$F_1 = +20$ N	$F_1 = +20$ N
$F_2 = +20$ N	$F_2 = -20$ N	$F_2 = -40$ N
$m = 5$ kg	$m = 5$ kg	$m = 5$ kg
$a =$ Unknown, solve	$a =$ Unknown, solve	$a =$ Unknown, solve

Calculate $\Sigma F = F_1 + F_2$.

$\Sigma F = (+20) + (+20)$	$\Sigma F = (+20) + (-20)$	$\Sigma F = (+20) + (-40)$
$\Sigma F = +40$ N	$\Sigma F = 0$ N	$\Sigma F = -20$ N

Now select the equation $\Sigma F = ma$, isolate for the unknown value, fill in the known values, and solve:

$$a = \frac{\Sigma F}{m} \qquad\qquad a = \frac{\Sigma F}{m} \qquad\qquad a = \frac{\Sigma F}{m}$$

$$= \frac{+40 \text{ N}}{5 \text{ kg}} = +8 \text{ m/s/s} \qquad = \frac{+0 \text{ N}}{5 \text{ kg}} = 0 \text{ m/s/s} \qquad = \frac{-20 \text{ N}}{5 \text{ kg}} = -4 \text{ m/s/s}$$

In the first example, when both F_1 and F_2 are applied in the same direction, the net force is an unbalanced force of +40 N and causes an acceleration of +8 m/s/s. In the second example, the opposing forces of equal magnitude cancel each other out, and the net force of 0 N (no unbalanced force on the body) causes an acceleration of 0 m/s/s — the body does not change motion because no unbalanced force acts on it. In the third example, the opposing forces are not of equal magnitude, and the net force is an unbalanced force of –20 N and causes an acceleration of –4 m/s/s in the direction of the unbalanced force.

I used a separate step to sum the forces. Then I inserted the ΣF into the equation after isolating for a (acceleration). The separate summing can be skipped and the forces inserted into the equation after isolating for a to save a separate step.

Deriving the impulse–momentum relationship from the law of acceleration

Newton's second law is useful to explain the acceleration of a body at an instant in time. But performance isn't just about "instants" in time, and performance is not just about acceleration. When bodies interact, like a baseball bat hitting a baseball, or a player landing and then immediately jumping into the air again, the equal and opposite forces are applied the entire time, and the velocity of each body is affected for the entire period of time the forces are applied. Each body has a velocity at the start of the application of force, and it may have a different velocity when the force is no longer applied. I can rewrite Newton's second law, $\Sigma F = ma$, to show the effect on a body's velocity when an unbalanced force is applied over a time period. I'll substitute the equation for acceleration, $a = \dfrac{V_f - V_i}{\Delta t}$, into $\Sigma F = ma$.

Begin with $\Sigma F = ma$, and substitute $\dfrac{V_f - V_i}{\Delta t}$ for a. The equation becomes

$$\Sigma F = m \left(\frac{V_f - V_i}{\Delta t} \right)$$

Now multiply both the velocities by m and the equation becomes

$$\Sigma F = \frac{(mV_f - mV_i)}{\Delta t}$$

Multiplying both sides by Δt leads to

$$\Sigma F \Delta t = mV_f - mV_i$$

This equation is the impulse–momentum relationship. The left side of the equation, $\Sigma F \Delta t$, is the important quantity called *impulse*. Impulse is the product of force multiplied by the duration of time it is applied. The units of impulse are Newton · seconds, or N · s (the units of force multiplied by the units of time). Momentum, explained in Chapter 5, is the product of a body's mass and its velocity. The units of momentum are kilogram · meters/second, or kg · m/s (the units of mass multiplied by the units of velocity). The right side of the equation, $mV_f - mV_i$ (read as final momentum minus initial momentum), describes a change in the momentum of a body. The equation shows that an applied impulse can increase or decrease the momentum of the body over the time period.

A source of some confusion is how impulse, measured in N · s, can be equal to momentum, measured in kg · m/s. The equality exists because 1 Newton is the amount of force to accelerate 1 kg at 1 m/s/s, or 1 N = 1 kg · 1 m/s/s. So, N · s is the same as kg · m/s/s · s. When kg · m/s/s · s is reduced, the result is kg · m/s, the units for momentum. The units of measurement are equivalent on both sides of the equation.

The change in momentum can be rewritten as $m(V_f - V_i)$ when the mass of the body stays the same — this is the case for most movement, because body mass stays the same during the movement. When the mass stays the same, an impulse applied to a body causes the momentum to change because the body changes velocity — it speeds up or slows down in the direction of the impulse applied. The version of the impulse–momentum relationship written as $\Sigma F\Delta t = m(V_f - V_i)$ is the most useful version for human movement.

The impulse–momentum relationship shows that the combination of force and time (impulse) is what *causes* the change in motion of a body (an increase or decrease in momentum). In theory, the velocity of a body can be changed equally by a large force applied for a short time, or a small force applied for a long time. In fact, any combination of force and time that produces an equal impulse will cause the same change in the velocity of a body. The size of the force and the duration of the force application are manipulated by a performer depending on his ability and the rules of the game.

The *F* in *F*Δ*t* is usually the average force exerted on the body during the time period Δ*t*. In reality, the force increases in magnitude from 0 before contact to some maximum value, and then decreases back to 0 at the end of the period of contact. I explain more about the changing force magnitude in Chapter 17.

Newton's first law is also known as the law of conservation of momentum. The first law states that the velocity of a body remains constant if no unbalanced force acts on the body. If the velocity and the mass remain constant, then momentum stays the same, or momentum is conserved, when no unbalanced force acts on a body.

Applying the impulse–momentum relationship for movement analysis

All performance can be analyzed in accordance with the impulse–momentum relationship. All external forces on a body are applied for a period of time. The motion of the body, how fast and in what direction it travels after the external force is applied, is what a coach, a teacher, a fan, or anyone watching a performance sees and evaluates. When shooting a basketball, the player applies an impulse to the ball to try to get the ball moving fast enough and in the correct direction so the ball moves through the air to the basket. The impulse from the ground increases a player's upward momentum when she jumps into the air, and the impulse from the ground decreases a player's downward momentum when she lands. When a golfer strikes a golf ball, the impulse applied to the ball by the golf club changes the momentum of the ball so it is (hopefully) moving with the velocity needed for a successful shot.

In each case, the impulse applied to the body of interest (the basketball, the player jumping and landing, the golf ball) determines the velocity change of the constant mass of the body (the mass of the basketball, the mass of the player jumping and landing, the mass of the golf ball). If the velocity change is not correct — if the basketball shot is not successful, if the player does not jump high enough or perhaps travels sideways instead of up, if the golf ball does not travel far enough or misses the target — the problem with the motion stems from an improper impulse.

Because impulse is determined by the magnitude of the forces applied to the body and the duration of time the force is applied to the body, cues to improve performance must be intended to improve the magnitude and direction of the applied force and/or the length of time the force is applied. (In Chapter 16, I explain how to use impulse as the mechanical basis of performance evaluation.)

Newton's third law, the law of equal but opposite, applies to impulse as well as to force alone. When a jumper pushes on the ground for a period of time, the ground pushes on the jumper for the same period of time. The forces exerted by the ground on the jumper and by the jumper on the ground are equal and opposite. The time of force application is also the same (there can be no jumper force on the ground without ground force on the jumper). The impulses are of the same magnitude but in the opposite directions on the two bodies. This is sometimes quite evident. When a batter hits a baseball, the focus of the analysis is usually the change in momentum of the ball, and the impulse of the ball on the bat is not of much concern to anyone. But the impulse applied to the bat by the ball is evident when the bat shatters.

The impulse–momentum relationship is used to calculate the magnitude of the impulse that changes a body's velocity, and it's a step to calculate the average force exerted on the body during the time period. I'll demonstrate by calculating the impulse and average force exerted by a bat on a baseball when the batter hits a fastball travelling horizontally at 90 mph (this velocity is 40 m/s) and drives it in the opposite horizontal direction at 110 mph (49 m/s). A baseball has a mass of about 0.145 kg (about 5 ounces). The time of contact between the ball and bat is very brief, around 1/800th of a second, or 0.013 seconds. The following steps are a good way to solve for the impulse and the average force on the ball:

1. **Identify the descriptor(s) you're asked to solve for.**

   In this case, you're solving for impulse and average force.

2. **Set up a coordinate system for the direction of velocity and force.**

   In this case, I'll use + for away from the batter and – for toward the batter.

3. **Create a table of variables, filling in the known and unknown values.**

- Velocity of pitch (V_i) = –40 m/s (toward the batter)
- Velocity of ball after bat contact (V_f) = +49 m/s (away from the batter)
- Mass (m) = 0.145 kg
- Δt = 0.013 s
- $\Sigma F \Delta t$ = impulse from the bat; unknown, to be solved
- Force (F) = average force from the bat; unknown, to be solved

4. **Select the impulse–momentum relationship, recognize that the unknown Ft is already isolated, fill in the known values, and solve.**

$$Ft = m(V_f - V_i) = 0.145 \text{ kg } ([+49 \text{ m/s}] - [-40 \text{ m/s}])$$

$$= 0.145 \text{ kg } (+89 \text{ m/s}) = +12.9 \text{ kg} \cdot \text{m/s (or} +12.9 \text{ N} \cdot \text{s)}$$

The bat applies +12.9 N · s of impulse to the ball during contact. (The ball, of course, applies –12.9 N · s of impulse to the bat during contact.)

5. **Using the definition of impulse = $F\Delta t$ = 12.9 N · s, isolate and solve for F.**

$$F = \frac{+12.9 \text{ N} \cdot \text{s}}{\Delta t} = \frac{12.9 \text{ N} \cdot \text{s}}{0.013 \text{ s}} = +992.3 \text{ N}$$

The average force applied to the baseball by the bat is +992.3 N, or 992.3 N applied opposite to the original direction of the baseball's velocity.

The direction of the average force applied is opposite to the initial velocity of the baseball at contact. The positive force slows down and stops the velocity of the ball in the negative direction (reduces the ball's motion to 0) and then speeds it up in the positive direction.

In the baseball example, the original velocity and the final velocity were in the horizontal direction. The only external force on the ball was the force applied by the bat. When using the impulse–momentum relationship to solve for the magnitude of the impulse and the average force from the floor applied to a body moving in the vertical direction, the same steps are followed, but the weight of the body must also be considered. I'll demonstrate by calculating the impulse and average force exerted by the floor on a 50 kg volleyball player landing from a vertical jump and immediately jumping straight up. The player is traveling downward at 2 m/s at initial floor contact and is traveling upward at 3 m/s after contacting the floor for 0.4 s.

The following steps show how the player's weight needs to be included in the calculations to solve for the impulse and the average force exerted by the floor:

1. **Identify the descriptor or descriptors you're asked to solve for.**

   In this case, you're solving for the impulse from the floor and average force from the floor.

2. **Set up a coordinate system for the direction of velocity and force.**

   In this case, I'll use + for upward and – for downward.

3. **Create a table of variables, filling in the known and unknown values.**

   - Velocity at initial floor contact (V_i) = –2 m/s (downward)
   - Velocity at end of floor contact (V_f) = +3 m/s (upward)
   - Mass (m) = 50 kg
   - $\Delta t = 0.4$ s
   - Weight = 50 kg · –10 m/s/s = –500 N (weight always acts downward)
   - $F\Delta t$ = impulse from the floor; unknown, to be solved
   - Force (F) = average force from the floor; unknown, to be solved

4. **Select the impulse–momentum relationship, recognize that the unknown $F\Delta t$ is already isolated, fill in the known values, and solve.**

   $$Ft = m(V_f - V_i) = 50 \text{ kg } ([+3 \text{ m/s}] - [-2 \text{ m/s}])$$
   $$= 50 \text{ kg}(+5 \text{ m/s}) = +250 \text{ kg} \cdot \text{m/s (or 250 N} \cdot \text{s)}$$

   When working in the horizontal direction with only one force on the body, the impulse calculated is produced by that single force. But in the vertical direction, the weight of the body *always* acts, and weight must be taken into consideration along with any other external force acting on the body in the vertical direction. During any time period, weight creates a downward impulse on the body in the vertical direction. The downward force of gravity must be included when calculating the average force.

5. **Using the definition of impulse $= F\Delta t = 250$ N · s, isolate and solve for F.**

   $$F = \frac{+250 \text{ N} \cdot \text{s}}{\Delta t} = \frac{+250 \text{ N} \cdot \text{s}}{0.4 \text{ s}} = +1,000 \text{ N}$$

At this point, recognize that the F solved for is the magnitude of the unbalanced force on the volleyball player. To solve for the force from the floor, solve for the floor force by calculating the sum of the forces on the player.

Recognize that ΣF is the sum of the two forces, the player's weight and the upward vertical force from the floor.

Solve for the net force in the vertical using $\Sigma F_{\text{Direction}}$. The steps are to write out the equation including the forces in the vertical direction, isolate for the unknown force, insert the magnitude and direction of the known values in the equation, and solve the equation:

$$\Sigma F_{\text{Vertical}} = \text{Weight} + \text{Floor Force}$$

Now isolate for floor force:

$$\text{Floor Force} = \Sigma F - \text{Weight}$$

$$\text{Floor Force} = (+1{,}000 \text{ N}) - (-500 \text{ N}) = +1{,}500 \text{ N}$$

The average force applied to the player by the floor is +1,500 N, or 1,500 N applied opposite to the original direction of the player's downward velocity. The +1,500 N force includes the force from the floor to offset the player's weight (+500 N) and the force from the floor to cause the change in motion of the player (+1,000 N).

Chapter 7

Looking At Force and Motion Another Way: Work, Energy, and Power

. . . .

In This Chapter

▶ Measuring work

▶ Mechanizing energy in its kinetic and potential forms

▶ Powering up positive and negative power

▶ Conserving mechanical energy

▶ Relating work and energy

. . . .

A *force* is a push or a pull exerted by one body on another body. Newton's laws of motion (explained in Chapter 6) provide one way to explain the effect of a force on a body. Another way to look at force acting on a body is with the ideas of work, energy, and power. These three terms are used in everyday conversation in a variety of ways: "I have a lot of work to do," "Work takes up a lot of my day," or "He really worked hard to finish on time"; "The party will be a high-energy event," "My house uses a lot of energy," or "I spent a lot of energy thinking through the problem"; and "the power of the people," "the power of love," or "She runs with a lot of power." While the use of the terms in everyday conversation is often not far from the meaning of the term in mechanics, understanding the mechanical meaning of each term is critical for precision.

In this chapter, I begin with the mechanical definition of *work*. I explain the term *energy* as the capacity of a body to do work, and *power* as the rate at which work is performed. (It's not the same as how quickly homework gets completed!) I cover the conservation of mechanical energy and the work–energy relationship to show how work and energy provide another way to explain the motion of bodies.

Working with Force

When an external force is applied to a body and the body moves, mechanical work is performed. *Work* is defined as the product of an external force and the displacement of the body. Mathematically, that's Work = Force (F) × Displacement (p), or Work = Fp. The unit of work is the Newton-meter (Nm), reflecting the unit of force, the Newton (N) multiplied by the unit of displacement, the meter (m). But because the unit Nm is also the unit of *torque* (the turning effect of a force; see Chapter 8), the unit used to describe work is the Joule: 1 Joule = 1 Nm of work.

Work is done by each external force acting on a body. For example, when you lift a box, the external force you apply does work on the body, and the external force applied by gravity does work on the body. Generally, the focus is on the work produced by *one* of the external forces applied to the body being moved.

Figure 7-1 shows an employee lifting a 100 N box from a table, holding it in position while a label is applied, and then lowering the box back to the table. The box gets raised 0.2 m above the tabletop. The average force applied by the employee is 100 N while raising the box (the ascent phase), while holding the box in position to be labeled (the holding phase), and while lowering the box (the descent phase).

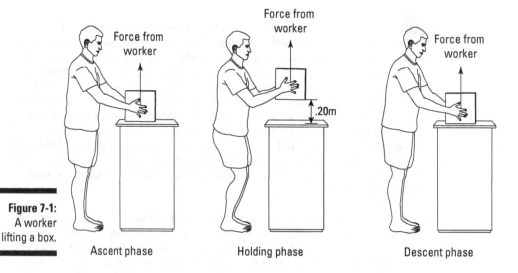

Figure 7-1: A worker lifting a box.

Ascent phase   Holding phase   Descent phase

I use the word *average* to describe the force because the external force applied by the employee to the box varies during both the ascent and descent phases of moving the box. As explained in Chapter 6, an unbalanced external force is required to change the motion of a body. The box is at rest on the tabletop when the employee begins the task, again when it's held by the employee as the label is applied, and yet again when it's set on the tabletop by the employee at the end of the task.

To start the box moving upward, there must be an unbalanced external force upward on the box. The employee must apply an upward force greater than the magnitude of the weight of the box to create the unbalanced force upward. To stop the box moving upward, there must be unbalanced external force acting downward on the box. The employee must apply an upward force less than the magnitude of the weight of the box to create the unbalanced downward force.

To hold the box in position, the employee must apply an upward force equal in magnitude to the weight of the box so there is no unbalanced external force acting on the box.

To start the box moving downward there must be an unbalanced force downward on the box. The employee must apply an upward force less than the magnitude of the weight of the box to create the unbalanced force downward. To stop the box moving downward, there must be an unbalanced force acting upward on the box. The employee must apply an upward force greater than the magnitude of the weight of the box to create the unbalanced upward force.

Over the entire task, the magnitude of the average force applied upward by the employee is equal in magnitude to the weight of the box.

Calculating the mechanical work exerted on a body by an external force requires description of the external force (average magnitude and direction) and the displacement of the body (how far and in what direction). I'll show the steps to calculate the work done on the box by the employee during the three phases of the task: the ascent phase, the holding phase, and the descent phase.

Ascent phase: Calculate the mechanical work on the box by the employee.

1. **Create a table of variables, filling in the known and unknown values.**

   - Average force (F) = +100 N (the employee only pulls up on the box)

   - Displacement (p) = +0.2 m (the box rises during the ascent phase)

2. Select the equation for work, fill in the known values, and solve.

Work = Force · Displacement = (+100 N) · (+0.2 m) = +20 Nm = +20 J

The employee did +20 J of mechanical work on the box during the ascent phase.

Holding phase: Calculate the mechanical work on the box by the employee.

1. Create a table of variables, filling in the known and unknown values.

- Average force (F) = +100 N (the employee only pulls up on the box)
- Displacement (p) = +0.0 m (the box does not move during this phase)

2. Select the equation for work, fill in the known values, and solve.

Work = Force · Displacement = (+100 N) · (0.0 m) = 0 Nm = 0 J

The employee did 0 J of mechanical work on the box during the holding phase.

Descent phase: Calculate the mechanical work on the box by the employee.

1. Create a table of variables, filling in the known and unknown values.

- Average force (F) = +100 N (the employee only pulls up on the box)
- Displacement (p) = –0.2 m (the box lowers during the descent phase)

2. Select the equation for work, fill in the known values, and solve.

Work = Force · Displacement = (+100N) · (–0.2m) = –20 Nm = –20 J

The employee did –20 J of mechanical work on the box during the descent phase.

When the direction of the applied external force and the direction of the displacement are the same, + mechanical work is done on the body. When the direction of the applied external force and the direction of the displacement are the opposite, – mechanical work is done on the body.

When an external force is applied to a body at rest, no mechanical work is done on the body. A body must move for an external force to perform mechanical work on the body.

Energizing Movement

Mechanical energy is defined as the capacity of a body to do work. The unit of energy is the Joule. In biomechanics, two types of mechanical energy get the most attention:

✔ **Kinetic energy:** Kinetic energy is present in a moving body. If a dropped bowling ball lands on your foot, the energy in the moving mass will have an effect on your foot. (It will do work on your foot as it lands.)

✔ **Potential energy:** Potential energy is present in a body because of its position or because it's been deformed. A bowling ball lying at rest on your foot is of little consequence, but a bowling ball raised and held above your foot is a concern because it has the potential to start moving if released. Holding a rubber band unstretched between your hands and releasing one end causes no pain, but if it's stretched (deformed) before release, it'll snap back and sting your fingers.

In this section, I explain kinetic energy and the two forms of potential energy.

Kinetic energy

Kinetic energy (*KE*) is defined as the energy present in a body because it's moving. Kinetic energy depends on the mass of the body and its velocity squared, as described in the equation $KE = \frac{1}{2}mv^2$. A more massive body has greater kinetic energy, and a faster moving body has greater kinetic energy, but the velocity of the body has the greater effect on kinetic energy because velocity is squared in the equation (v^2). I'll show the greater effect of velocity by calculating the increase in *KE* when mass and velocity are increased. The measurement unit of kinetic energy is the Joule (J).

Table 7-1 shows the effect of increased mass on kinetic energy. Player 2 has a mass of 120 kg, 1.5 times, or 50 percent greater than, the 80 kg mass of player 1. When they both run at 8 m/s, the *KE* of player 2 (3,840 J) is 1.5 times, or 50 percent greater than, the *KE* of player 1 (2,560 J). The *KE* of player 2 (8,640 J) is also 1.5 times, or 50 percent greater than, player 1 (5,760 J), if they both run at 12 m/s. The increase in *KE* is proportional to the increase in the mass of the body; a 1.5 times, or 50 percent, increase in mass results in a 1.5 times, or 50 percent, increase in *KE*.

Table 7-1			Effect of Increasing Mass and Velocity on *KE*	
	Mass	*Velocity*	*Calculation of KE = $\frac{1}{2}mv^2$*	*KE*
1	80 kg	8 m/s	$= \frac{1}{2}$ (80 kg)(8 m/s)$^2 = \frac{1}{2}$ (80 kg)(64 m^2/s^2)	2,560 J
		12 m/s	$= \frac{1}{2}$ (80 kg)(12 m/s)$^2 = \frac{1}{2}$ (80 kg)(144 m^2/s^2)	5,760 J
2	120 kg	8 m/s	$= \frac{1}{2}$ (120 kg)(8 m/s)$^2 = \frac{1}{2}$ (120 kg)(64 m^2/s^2)	3,840 J
		12 m/s	$= \frac{1}{2}$ (120 kg)(12 m/s)$^2 = \frac{1}{2}$ (120 kg)(144 m^2/s^2)	8,640 J

Emilie du Chatelet

Emilie du Chatelet (1706–1749) was a daughter of French nobility. Disregarding the practice of educating women at the time, du Chatelet's parents acknowledged their daughter's intellectual curiosity and brought in tutors to teach her math and science in addition to the society-accepted education in literature, music, and dance. She is recognized for translating Newton's *Principia* to French (more than 250 years later, hers is still considered the standard French translation), but du Chatelet is notable for combining existing theoretical and practical experiments to show that the kinetic energy of a body ($\frac{1}{2}mv^2$) is different from the momentum of a body (*mv*). Because of the male domination of science at the time, her contributions were overlooked for almost 200 years. But du Chatelet is beginning to receive greater recognition for her scientific contributions.

Table 7-1 also shows the effect of increased velocity on kinetic energy. When player 1 increases velocity from 8 to 12 m/s, a 1.5 times, or 50 percent, increase in velocity, the *KE* increases by 2.25 times, or 125 percent, more than doubling from 2,560 J to 5,760 J. A similar 1.5 times, or 125 percent, increase is seen when player 2 increases velocity 1.5 times from 8 m/s (3,840 J) to 12 m/s (8,640 J). The increase in *KE* is proportional to the square of the increase in velocity (1.5 times increase in velocity = 2.25 times increase in *KE*, [$1.5^2 = 2.25$] because the velocity contribution to *KE* is squared [v^2]).

KE is the energy because a body is moving. The unit for *KE* is the Joule, but mass (kg) and velocity (m/s) are the contributors to *KE*. It's not obvious how the contributing units of kg and m/s result in a Joule, but using a little mathematical gymnastics on the equation shows $KE = \frac{1}{2}mv^2 = (\text{kg}) \cdot (\text{m/s})^2 = \text{kg} \cdot (\text{m} \cdot \text{m/s/s}) = \text{kg} \cdot \text{m} \cdot \text{m/s/s}$. By regrouping the terms, this becomes $\text{kg} \cdot \text{m/s/s} \cdot \text{m}$, and selectively adding parentheses shows this to be $(\text{kg} \cdot \text{m/s/s}) \cdot \text{m}$, which reduces to Nm = J. *Voilá!*

Potential energy

Potential means something is possible. *Potential energy* means a body has potential kinetic energy. The potential kinetic energy of the body will become kinetic energy if certain conditions of the body change. In this section, I explain two forms of potential energy: gravitational potential energy (potential energy related to the position of a body) and strain energy (potential energy related to the deformation of a body). The measurement unit of both forms of potential energy is the Joule (J).

Gravitational potential energy

Gravitational potential energy (GPE) is the potential energy in a body because of its position relative to a reference point below it. When a body is positioned above the ground, as it starts to fall toward the ground, the gravitational acceleration of –10 m/s/s will increase its downward velocity (and kinetic energy). The gravitational potential energy in a body is related to its height (h; actually, the displacement the body will *fall* to a reference point), its mass (m), and the gravitational acceleration ($g = –10$ m/s/s) by the equation $GPE = mgh$. The higher a body is above the reference point, the greater its *GPE*.

Figure 7-2 shows a mountain climber on a rock face. The climber has a mass of 70 kg, and is at a height 100 m above the ground but also 50 m above a ledge jutting out from the rock face. In the case of a fall, the climber could either hit the ledge or fall all the way to the ground. I'll show the steps to calculate the *GPE* in the climber relative to both the ground and the projecting ledge.

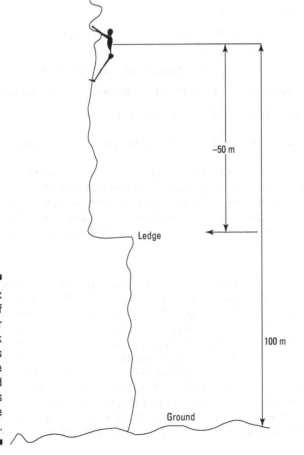

–50 m

Ledge

100 m

Ground

Figure 7-2: The *GPE* of a climber on a rock face differs if the ledge or ground is used as a reference point.

Fall to the ground: Calculate the *GPE*.

1. **Create a table of variables, filling in the known and unknown values.**

   - Climber mass (m) = 70 kg
   - Gravitational acceleration (g) = –10 m/s/s (the acceleration from gravity is always *down*)
   - Height (h) = –100 m (the climber will fall *down* 100 m)

2. **Select the equation for *GPE*, fill in the known values, and solve.**

   $GPE = mgh = $ (70 kg) · (–10 m/s/s) · (–100 m) = 70,000 J

The climber has 70,000 J of *GPE* relative to the ground 100 m below.

Fall to the ledge: Calculate the *GPE*.

1. **Create a table of variables, filling in the known and unknown values.**

   - Climber mass (m) = 70 kg
   - Gravitational acceleration (g) = –10 m/s/s (the acceleration from gravity is always *down*)
   - Height (h) = –50 m (the climber will fall *down* 50 m)

2. **Select the equation for *GPE*, fill in the known values, and solve.**

   $GPE = mgh = $ (70 kg) · (–10 m/s/s) · (–50 m) = 35,000 J

The climber has 35,000 J of *GPE* relative to the ledge 50 m below.

A body may have *GPE* even when resting on the ground. The *GPE* is present because the center of mass of the body is above the contact point with the ground. The center of mass of the human body (see Chapter 8) is located at about the level of the navel, or belly button. The presence of *GPE* in an upright body becomes evident when a person faints and falls to the ground.

Strain energy

Strain energy (*SE*) is the potential energy in a body because of a deformation or change in shape. *Strain* is the mechanical term for deformation (strain is explained in more detail in Chapter 12). When a body undergoes strain, energy is stored in the deformed material of the body. The strain energy in a particular body is described by the equation $SE = \frac{1}{2}k\Delta x^2$, where Δx is the displacement, in meters, during deformation and k is the stiffness, or resistance to deformation, of a body measured in N/m (more on stiffness in Chapter 12). The equation shows that *SE* increases with greater deformation (as Δx^2 increases) and/or as k increases (the body has a greater resistance to deformation). However, there is a limit to how much deformation a specific material can undergo before it breaks (fractures, tears, snaps, whatever damage to that specific material is called), so it isn't simply a matter of deforming a body more to store more energy in it. Measuring *SE* is more difficult than measuring either *KE* or *GPE*.

Although difficult to quantify, strain energy is critical to human performance. For example, the pole used in pole vaulting is intended to temporarily store *SE* when deformed by the athlete (as shown in Figure 7-3), before regaining shape and pushing the vaulter up to (and hopefully over) the bar. A modern vaulting pole is made of high-tech materials and is carefully matched to the physical characteristics and skill level of the vaulter (at least at the elite levels — novice pole vaulters often make do with what's in the track-and-field storage shed). The storage of strain energy is also considered in the modern design of baseball bats, hockey sticks, and other implements used in sports.

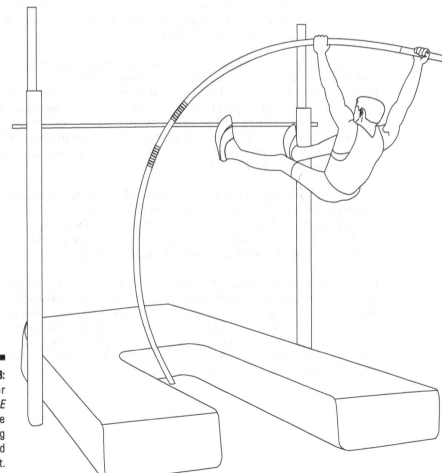

Figure 7-3:
The vaulter stores *SE* in the pole by planting it and bending it.

The human body also uses *SE* stored in soft tissues of the body, especially muscle and tendon, to move efficiently. In Chapter 12, I explain the use of strain energy (or as it's sometimes called, *elastic energy*) by muscles and tendons in more detail.

Conserving Mechanical Energy

An overriding principle of the universe is that no new energy is being created. In other words, energy is conserved, and can only be converted from one form to another. Energy can be converted between mechanical, chemical, heat, sound, and potential energy, but the sum of all the energy forms remains the same. In biomechanics, the conservation of total mechanical energy (total mechanical energy = TME = *KE* + *GPE*) is of interest. I'll use the example of a 40-kg gymnast jumping straight up and coming straight down to explain the conservation of mechanical energy.

The top part of Figure 7-4 shows a gymnast jumping straight up into the air (the *rising phase*) to a peak height and coming straight back down (the *falling phase*). Although the path of the gymnast is only vertical, I've spread the vertical position of the gymnast horizontally so the vertical positions match with the graph of *KE* and *GPE* shown in the bottom part of Figure 7-4. The bottom part shows TME as a constant value during the jump, with decreases in *KE* offset by increases in *GPE* during the rising phase and increases in *KE* offset by decreases in *GPE* during the falling phase.

At take-off, the vertical velocity is 6.0 m/s and the center of mass is 0.8 m above the ground. The *KE* of the gymnast is 720 J ($\frac{1}{2}$ · 40 kg · [+6.0 m/s]^2) and the *GPE* is 320 J (40 kg · [−10 m/s/s] · [−0.8 m]). The TME = *KE* + *GPE* = 720 J + 320 J = 1,040 J. The conservation of mechanical energy says the TME remains constant at 1,040 J during the gymnast's vertical projectile motion.

On the way up to peak height, the decrease in *KE* as the upward vertical velocity decreases is offset by the increase in *GPE* as the gymnast rises above the ground. At peak height, the gymnast is 1.8 m above the ground and the vertical velocity is 0 m/s (the gymnast is no longer moving upward but hasn't yet started to fall back to the ground). At peak height, *KE* = $\frac{1}{2}$ (40 kg)(0 m/s)^2 = 0 J, but the *GPE* = (40 kg)(−10 m/s/s)(−1.8 m) = 1,040 J. All the *KE* has been converted to *GPE*, and TME is the same as at take-off.

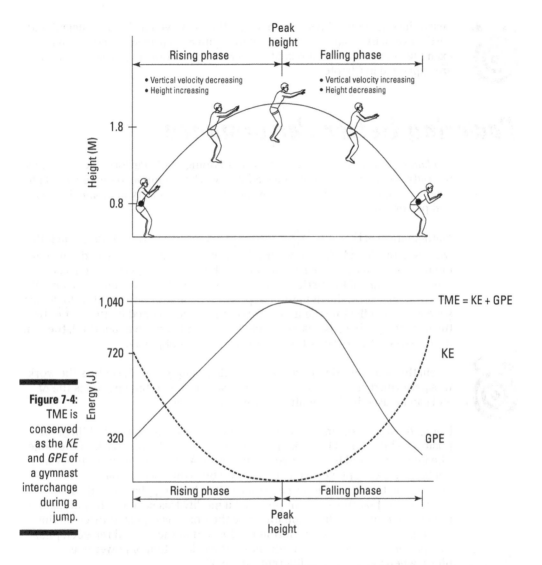

Figure 7-4: TME is conserved as the *KE* and *GPE* of a gymnast interchange during a jump.

On the way back down, the increase in *KE* as the downward vertical velocity increases is offset by the decrease in *GPE* as the gymnast gets closer to the ground. At ground contact, the gymnast's vertical velocity is –6.0 m/s and the center of mass is back to 0.8 m above the ground. The *KE* of the gymnast is 720 J ($\frac{1}{2}$ · 40 kg · [–6.0 m/s]^2) and the *GPE* is 320 J (40 kg · [–10 m/s/s] · [–0.8 m]). The TME = *KE* + *GPE* = 720 J + 320 J = 1,040 J. The TME remains constant — it is conserved — during the entire jump. Changes in *KE* are offset by changes in *GPE* as velocity and height change.

I used the equations of constant acceleration to solve for the peak height and vertical velocity at ground contact of the gymnast's projectile motion in the example above. The equations of constant acceleration and projectile motion are explained in Chapter 5.

Powering Better Performance

Mechanical power is defined as the rate of doing work. Performance depends not only on being able to produce a lot of work (*Fp*) but also on being able to produce the work quickly. Power is an important quantity in biomechanics and performance.

Mathematically, Power $= \dfrac{\text{Work}}{\Delta t}$, where Δt is the time period during which the work was produced. The time period Δt is calculated as $t_f - t_i$, (read time final minus time initial), where t_f is the time at the end of the period and t_i is the time at the start of the period. Dividing "work" by "time period" measures the rate of doing work, the definition of power. The unit of power is the Joule per second (J/s), reflecting the unit of force, the Joule, divided by the unit of time, the second. But because saying Joule per second is quite a mouthful, the unit used to describe power is the watt: 1 watt = 1 J/s of power.

Often, the time period to perform the work is simply stated, such as the work was performed in 0.25 seconds. In this case, the stated time period is the Δt and can be used in the equation to calculate power.

Figure 7-1 shows an employee lifting a 100 N box from a table, holding it in position while a label is applied, and then lowering the box back to the table. The box was raised 0.2 above the tabletop. An average force of 100 N was applied by the employee while raising the box (the ascent phase), while holding the box in position to be labeled (the holding phase), and while lowering the box (the descent phase). I'll modify this scenario by adding that the task of lifting, holding, and lowering the box was repeated twice. In the first rep, each phase was performed in 1 s, and in the second rep each phase was performed in 0.5 s. Table 7-2 shows the calculation of power in each phase when performed in different time periods.

Table 7-2   Power Calculated on Two Repetitions of a Lifting Task

Rep	Phase	Force	Displacement (p)	Work (Fp)	Time (Δt)	Power (Work/Δt)
	Ascent	+100 N	+0.2 m	+200 J	1.0 s	+200 W
First	Holding	+100 N	0 m	0 J	1.0 s	0 W
	Descent	+100 N	−0.2 m	−200 J	1.0 s	−200 W

Rep	Phase	Force	Displacement (p)	Work (Fp)	Time (Δt)	Power (Work/Δt)
	Ascent	+100 N	+0.2 m	+200 J	0.5 s	+400 W
Second	Holding	+100 N	0 m	0 J	0.5 s	0 W
	Descent	+100 N	−0.2 m	−200 J	0.5 s	−400 W

Table 7-2 shows the work performed during comparable phases is the same regardless of the time taken to perform the work during the phase. During the ascent phase, +200 J of work was performed on the box whether it was lifted in 1 s or 0.5 s. The power output was affected by the time taken. When the ascent phase was performed in 1 s, the power output was +200 W; when the phase was performed in 0.5 s, the power output was doubled to +400 W. The same effect occurred during the descent phase when the negative power output was affected by the time to complete the phase.

Higher power is produced when work is completed in a shorter time period. Performing the work in half the time reflects a doubling of the power output. Power output is inversely proportional to the time to complete the work.

Work is defined as Force · Displacement, and *power* is defined as Work/Δt. Substituting Force · displacement for work in the equation for power gives $\text{Power} = \frac{(\text{Force} \cdot \text{Displacement})}{\Delta t}$. Velocity (see Chapter 5) is defined as $\frac{\text{Displacement}}{\Delta t}$, so Power = Force · Velocity is an alternate equation for power.

A powerful person quickly produces work. Linking the concept of power to the term *acceleration* (explained in Chapter 5) shows that a high power means a high acceleration. Sport analysts, or color commentators, often describe a good player as being able to "accelerate with great power." Biomechanically, the analyst is redundant, because power describes doing work quickly and acceleration means changing velocity quickly.

The Work–Energy Relationship

As the definitions suggest, *work* and *energy* are related. *Work* is defined as Fp (force multiplied by displacement), and *energy* is defined as the capacity of a body to do work. The work–energy relationship, as the name implies, relates the amount of work done on a body to the change in the total mechanical energy in the body. Mechanical energy is one form of energy, and all three types of mechanical energy are important in biomechanics: kinetic (*KE*), gravitational potential energy (*GPE*), and strain energy (*SE*). The work–energy relationship states that the total mechanical energy in a body changes only when work is applied to the body by an external force other than gravity.

When gravity is the only external force acting on a body, the total mechanical energy in the body remains constant, but the type of mechanical energy can be converted between *KE* and *GPE* as explained with the gymnast jumping up and down.

The work–energy relationship is $Fp = \Delta KE + \Delta GPE$. The force ($F$) in the equation is any external force other than gravity. The Greek letter Δ (delta) means "change in," so ΔKE and ΔGPE mean "change in" the *KE* and *GPE*. That is, the change in energy from the start of the application of work to the end of the application of work (Δenergy = energy$_{final}$ – energy$_{initial}$).

I'll explain the work–energy relationship using the example of the box raised and lowered from a table by an employee. This example shows how the external work performed by the employee on the box changes the *GPE* of the box. (I describe the task in detail earlier in this chapter.) In Figure 7-5, I've redrawn the ascent (lifting) and descent (lowering) phases of the task. I've added that the center of mass of the box is 0.1 m above the table surface when the box is on the tabletop.

Box at end of ascent

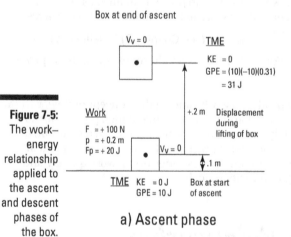

Figure 7-5:
The work–
energy
relationship
applied to
the ascent
and descent
phases of
the box.

a) Ascent phase

Box at start of descent

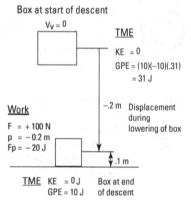

b) Descent phase

The employee first lifts the box above the table surface during the ascent phase. The box begins the ascent phase at rest (velocity = 0) on the surface of the table. With no velocity, the initial *KE* = 0. Using the table surface as the reference, the initial *GPE* = (10 kg) · (–10 m/s/s) · (–0.1 m) = 10 J. During the ascent phase, the employee applies an average vertical force of +100 N to the 10 kg box. At the end of the ascent phase, the box is at rest (velocity = 0) and the center of mass is 0.3 m above the surface of the table; the center of mass has undergone a displacement of +0.2 m. With no velocity, the final $KE = 0$. The final $GPE = mgh = (10 \text{ kg}) \cdot (-10 \text{ m/s/s}) \cdot (-0.3 \text{ m}) = 30$ J. Applying the work–energy relationship $Fp = \Delta KE + \Delta GPE$:

$$Fp = (KE_{final} - KE_{initial}) + (GPE_{final} - GPE_{initial})$$

Substituting in the values:

$$+100 \text{ N} \cdot +0.2 \text{ m} = (0 \text{ J} - 0 \text{ J}) + (20 \text{ J} - 10 \text{ J})$$

$$+20 \text{ J} = +20 \text{ J}$$

The +20 J of work performed by the employee on the box acted to increase the *GPE* of the box by +20 J. When an external force other than gravity performs positive work on a body, the mechanical energy in the body increases.

In the descent phase, the employee lowers the box to the table surface. The box begins the descent phase at rest (velocity = 0) with the center of mass 0.3 m above the surface of the table. With no velocity, the initial $KE = 0$. Using the table surface as the reference, the initial $GPE = (10 \text{ kg}) \cdot (-10 \text{ m/s/s}) \cdot (-0.3 \text{ m}) = 30$ J. During the descent phase, the employee applies an average vertical force of +100 N to the 10 kg box. At the end of the descent phase, the box is at rest (velocity = 0) on the tabletop with the center of mass now 0.1 m above the surface of the table; the center of mass has undergone a displacement of –0.2 m. With no velocity, the final $KE = 0$. The final $GPE = mgh = (10 \text{ kg}) \cdot (-10 \text{ m/s/s}) \cdot (-0.1 \text{ m}) = 10$ J. Applying the work–energy relationship $Fp = \Delta KE + \Delta GPE$:

$$Fp = (KE_{final} - KE_{initial}) + (GPE_{final} - GPE_{initial})$$

Substituting in the values:

$$+100 \text{ N} \cdot -0.2 \text{ m} = (0 \text{ J} - 0 \text{ J}) + (10 \text{ J} - 30 \text{ J})$$

$$-20 \text{ J} = -20 \text{ J}$$

The –20 J of work performed by the employee on the box acted to decrease the *GPE* of the box by –20 J. When an external force other than gravity performs negative work on a body, the mechanical energy in the body decreases.

One form of mechanical energy is *GPE,* which is present because the force of gravity, or weight (mg), acts on the body and can change the kinetic energy of the body. When the force of gravity acts alone on a body, *GPE* will be converted to *KE,* or vice versa, but gravity by itself cannot increase or decrease the total mechanical energy in the body. This is because gravity would show up on both sides of the equation describing the work–energy relationship, as the force applied over the displacement and as the cause of the *GPE*. So the work–energy relationship is best stated as "the work done by an external force *other than gravity* acting on a body causes a change in the total mechanical energy in the body."

Consider the mountain climber from the example above. When the 70 kg climber is 100 m above the ground but not moving, the total mechanical energy in the body is $GPE = (40 \text{ kg}) \cdot (-10 \text{ m/s/s}) \cdot (-100 \text{ m}) = 70,000 \text{ J}$. If the climber falls the 100 m to the ground, downward velocity increases on the way down and KE increases, but GPE decreases because the climber is closer to the ground, the reference point. At contact with the ground, there is no GPE ($h = 0$) and all the energy in the climber has been converted to kinetic energy, so $KE = 70,000$. The total mechanical energy in the body is the same at the highest and lowest points, but the mechanical energy is all GPE at the highest point and all KE at the lowest point.

Because the amount of GPE before the climber falls is known (70,000 J), I can calculate the velocity of the 70 kg climber at ground contact:

> Known: $TE = 70,000 \text{ J} = KE + GPE$

At ground contact, all 70,000 J of mechanical energy is now KE because $GPE = (70 \text{ kg}) \cdot (-10 \text{ m/s/s}) \cdot (0 \text{ m}) = 0 \text{ J}$ (the h in mgh = 0). Using $KE = \frac{1}{2}mv^2$, isolating for v, the equation becomes

$$v = \sqrt{\frac{2KE}{m}}$$

Substitute in the known values and solve:

$$= \sqrt{\frac{2 \cdot 70,000}{70}} = \sqrt{\frac{140,000}{70}} = \sqrt{2,000} = \pm 44.7 \text{ m/s}$$

Because the climber is falling, the velocity of the climber at ground contact is –44.7 m/s (about 100 mph downward — it's no wonder falls from high heights are often fatal).

Energy comes in many forms, including mechanical, chemical, heat, light, and sound. No new energy is being produced in the universe. Although my example shows that work converts mechanical energy only between KE and GPE, in the grand scheme of things, energy converts among all the forms. For example, when a gymnast lands, the work of the upward force from the ground on the downward-moving gymnast performs negative work on the gymnast, reducing the KE of the gymnast by stopping the downward motion. But some of the KE is also converted to sound energy (the sound of the slap of the gymnast's feet on the ground), to heat energy (the soles of the gymnasts feet and the surface of the floor get a little bit warmer), and to strain energy (the floor mat and the tissues of the gymnast both deform during the landing).

Bungee jumping

Bungee jumping is a popular recreational activity among people willing to risk life and limb to the expected conversion of mechanical energy among *GPE*, *KE*, and, most important, *SE*.

In a bungee jump, a seemingly sane participant climbs to a high height (world record: 322 m!) where a rubber cord (the bungee cord) is secured around the jumper's ankle or waist. The jumper then leaves a perfectly safe position to fall headfirst toward the ground. The following figure shows the position and motion of a jumper at p_1, p_2, p_3, p_4, and p_5.

I'll explain the *GPE*, *KE*, and *SE* at each of the five positions during the fall:

✔ p_1: The jumper stands with no vertical velocity on the platform at maximum height (*h*) above the ground (the reference point for *GPE*). The bungee cord lies raveled on the platform. All the mechanical energy is

present as *GPE*. *KE* = 0 because vertical velocity (V_v) is 0, and *SE* = 0 because the bungee cord is not deformed.

✔ p_2: The fool, er, participant, has jumped off the platform (and may now be regretting it), and the downward V_v is increasing. The bungee cord is unraveling but is not yet stretching. The mechanical energy is present as *GPE* and *KE*. *GPE* is less than when the jumper was on the platform, because the jumper is getting closer to the ground ($h_2 < h_1$). *KE* has increased because the jumper's $-V_v$ has increased during the fall. *SE* = 0 because the bungee cord is unfurling but is not deformed.

✔ p_3: The jumper is still falling and has reached maximum downward V_v. The bungee cord is completely unfurled and is about to start stretching. The mechanical energy is

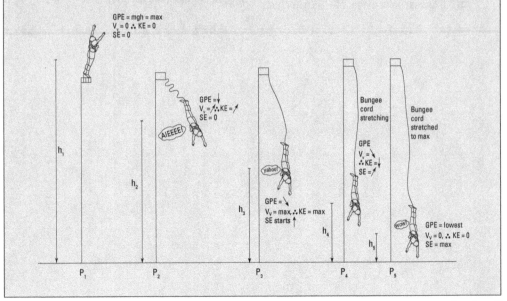

GPE = mgh = max
V_v = 0 ∴ KE = 0
SE = 0

GPE = ↓
V_v = ↑ ∴ KE = ↗
SE = 0

AIEEEE!

h_1

h_2

h_3

GPE = ↓
V_v = max, ∴ KE = max
SE starts ↑

yahoo!

Bungee cord stretching

Bungee cord stretched to max

GPE
V_v = ↓
∴ KE = ↓
SE = ↗

h_4

h_5

WOW!

GPE = lowest
V_v = 0, ∴ KE = 0
SE = max

P_1 P_2 P_3 P_4 P_5

(continued)

(continued)

present as *GPE* and *KE*. *GPE* is less than before because *h* continues to decrease. *KE* is now at its maximum value because downward V_v is at its maximum. *SE* = 0 because the bungee cord is completely unfurled but has not begun to stretch. Once the cord starts to stretch (deform), it will begin to pull upward on the jumper, decreasing V_v, and *SE* will begin to get stored in the cord.

✔ p_4: The person is still falling, but downward V_v is decreasing because the bungee cord is stretched and pulling upward on the downward-moving jumper. The mechanical energy is present as *GPE*, *KE*, and *SE*. *GPE* continues to decrease because the jumper continues to get closer to the ground. *KE* is less than it was at p_3 because the downward V_v is less than it was at p_3. *SE* is increasing as the cord is stretched (deforms).

✔ p_5: The downward V_v is 0, and the person is at the lowest point of the fall (*h* is minimum).

The bungee cord is stretched as much as it is going to stretch during the jump. The mechanical energy is present as *GPE* and *SE*. *GPE* is at its lowest value because *h* is minimum. *KE* = 0 because the V_v is 0. *SE* is at its maximum value because the cord is deformed as much as it is going to stretch. All the *KE* present has been converted to *SE*. While stretching, the cord performed negative work on the jumper to slow down and stop the fall.

Following p_5, the stretched bungee cords pulls the jumper upward. The strain energy in the cord is converted to *KE* (the upward V_v of the jumper increases) and to *GPE* (the jumper goes higher in the air). Typically, a jumper will go up and down several times before the mechanical energy is lost as heat and sound energy. The bungee jump is a success if the cord is not deformed to the point that the cord breaks (or as long as the stretching cord does not allow the jumper to hit the ground).

Part III
Investigating Angular Mechanics

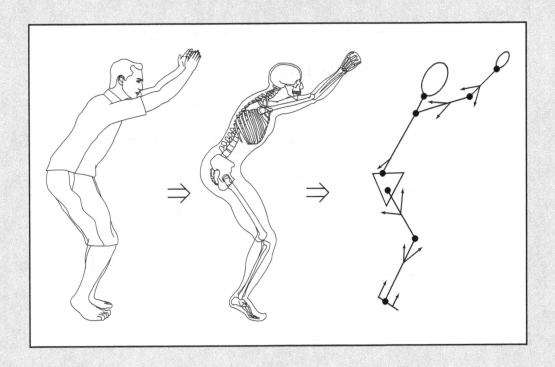

Find out more about calculating muscle force at the elbow joint in a free article at
http://www.dummies.com/extras/biomechanics.

In this part...

- Calculate torque as the product of a force and a moment arm.

- Explore the quantities of angular kinematics and their relation to linear kinematics.

- Apply a link segment model of the human body to show how muscle produces torque at a joint.

- Manipulate the moment of inertia to spin faster or slower.

- See how fluid mechanics explains why a ball curves in the air.

Chapter 8

Twisting and Turning: Torques and Moments of Force

C hanges in the *linear* motion of a body result from unbalanced *forces*. Changes in the *angular* motion of a body result from unbalanced *torques*. *Torque* is the turning effect of a force; it causes rotation. The four characteristics of a force are magnitude, direction, point of application, and line of action (see Chapter 4). The line of action of a force is an important characteristic in the description of the turning effect of a force. The line of action is the extension of the vector representing the force in both directions. The location of the line of action of a force relative to a specific axis determines if a force will tend to cause a change in the rotation of a body.

This chapter begins with an explanation of the mechanical concept of torque, the turning effect of a force. I show how the turning effect of a force at a specific axis is affected by the line of action of the force relative to the specified axis. Then I describe how a link segment model of the human body simplifies the complex anatomy of the human body to describe how a force causes rotation of a segment. Using the link segment model, I show how muscle serves as a torque generator and how other forces create a loading torque. I explain equilibrium as a condition of no unbalanced force or unbalanced torque acting on a body. Finally, I use the condition of equilibrium to identify the center of gravity of the human body. This chapter presents an overview of how force creates torque around an axis.

Defining Torque

Torque is the turning effect of a force, or the tendency of a force to cause rotation. The forearm rotating at the elbow and a ballet dancer turning en pointe are both examples of torque acting on a body. An unbalanced force tends to affect the linear motion of a body. An unbalanced force is determined by considering the magnitude and direction of the forces acting in a specific direction on a body. The line of action of a force is a critical characteristic in determining the effect of a force on a body. The line of action of a force is shown as a dashed line extending the force vector in both directions. A force applied to a body is identified as centric or eccentric depending on whether the line of action of the force passes through or beside the specified axis. I'll use Figure 8-1 to explain.

Figure 8-1:
A centric
force, an
eccentric
force, and a
force couple
acting on a
body.

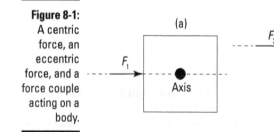 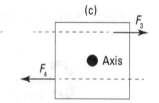

The body in Figure 8-1 is a box free to rotate around the axis indicated. Different forces are applied to the box. A force is called *centric* or *eccentric* depending on how the line of action of the force is positioned relative to a specified axis, or point around which the body can rotate. The line of action is the extension of the force vector in both directions and is represented with a dashed line.

A *centric* force occurs when the line of action of a force passes through the identified axis of rotation. *Centric* means "directed to the center," and a centric force tends to cause *translation* (linear motion) of the body. In Figure 8-1a, the centric force F_1 will cause the body to accelerate to the right because it represents an unbalanced horizontal force.

An *eccentric* force occurs when the line of action of a force does not pass through the identified axis of rotation. *Eccentric* means "directed off center," and an eccentric force tends to cause *translation* and *rotation* (angular motion) of the body. In Figure 8-1b, the eccentric force F_2 will cause the body to accelerate to the right because it represents an unbalanced horizontal force, but F_2 will also cause the body to rotate in the clockwise direction because the line of action does not pass through the axis of rotation. I refer to the torque created by F_2 as T_{F2}.

Two eccentric forces of equal magnitude applied to a body in opposite directions create a *force couple* (Figure 8-1c). A force couple tends to cause only *rotation* of the body. Because the forces are of equal magnitude but act in opposite directions, the linear effect of each force is canceled out ($\Sigma F = 0$) and no translation is caused. The lines of action of both F_3 and F_4 don't pass through the axis of rotation, and both forces tend to cause the body to rotate in the clockwise direction. The net torque on the box is the sum of the torques from F_3 and F_4 ($\Sigma T = T_{F3} + T_{F4}$).

Torque is the turning effect created by an eccentric force. How eccentric a force is — how off-center it is — is one factor affecting the torque created by a force.

Lining up for rotation: The moment arm of a force

Moment arm (MA) is the perpendicular (right-angle) distance from the line of action of a force to a specific axis of rotation. The moment arm measures the "eccentric-ness" of a force, in meters (m).

Figure 8-2 shows eccentric forces F_1, F_2, and F_3 applied to the box. Each force = 100 N. The length of the vector represents force magnitude; the tip, or point, of the arrow represents the force direction; and where the force touches the body is the point of application.

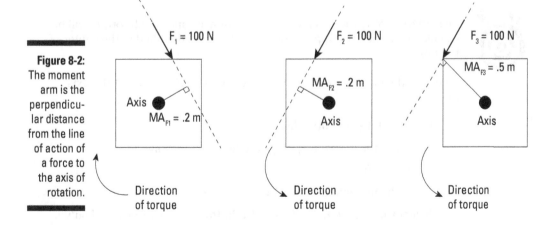

$F_1 = 100$ N

Axis

$MA_{F1} = .2$ m

Direction of torque

$F_2 = 100$ N

$MA_{F2} = .2$ m

Axis

Direction of torque

$F_3 = 100$ N

$MA_{F3} = .5$ m

Axis

Direction of torque

F_1 and F_2 are the same magnitude and have the same point of application, but the direction of the two forces is different. The line of action of each force is shown with the dashed line, and the MA of each force is shown as the perpendicular distance from the line of action of the force to the axis of rotation. The MA for both F_1 (MA_{F1}) and F_2 (MA_{F2}) is 0.2 m. F_1 will cause a torque in the clockwise direction, and F_2 will cause a torque in the counterclockwise direction.

F_3 has the same magnitude and direction as F_2, but F_3 has a different point of application. F_3 is applied at the corner of the box. The MA for F_3 shows that F_3 is more eccentric than F_2; the $MA_{F3} = 0.5$ m. F_3 will have a greater turning effect on the box than F_2, although both F_2 and F_3 will cause a torque in the counterclockwise direction.

For a given force, more torque is created with a longer moment arm than a shorter moment arm. For a given moment arm, more torque is created with a larger force than a smaller force.

Calculating the turning effect of a force

Torque (T) depends on the magnitude of the force (F) and the length of the moment arm (MA), stated as an equation $T = F \cdot MA$. Torque is a vector quantity, so both magnitude and direction of the turning effect must be specified. The units to express the magnitude of torque are the Nm, reflecting the unit of force, the Newton (N), multiplied by the unit of the moment arm, the meter (m). The direction of the torque must also be identified. The clockwise (CW) direction is the – direction, and the counterclockwise (CCW) direction is the + direction.

I'll demonstrate steps to calculate torque using F_1, F_2, and F_3 and the moment arms shown in Figure 8-2, starting with the torque created by F_1.

Be sure to identify the *specific* axis of rotation around which torque will be calculated. In these examples, the axis is the axis identified in the center of the box.

Calculate the magnitude of the torque created by $F_1 = 100$ N at the axis indicated:

1. **Create a table of variables.**

   - The torque from F_1 (T_{F1}): Unknown, to be solved

   - $F_1 = 100$ N

   - The moment arm for F_1 (MA_{F1}) = 0.2 m

2. **Select the equation $T = F \cdot MA$, fill in the known values, and solve.**

   $$T_{F1} = F_1 \cdot MA_{F1} = (100 \text{ N}) \cdot (0.2 \text{ m}) = 20 \text{ Nm}$$

Identify the direction of the torque:

> From Figure 8-2, F_1 will cause the body to rotate in the clockwise, or –, direction. The torque created by F_1 around the axis is equal to –20 Nm.

Calculate the magnitude of the torque created by $F_2 = 100$ N at the axis indicated:

1. **Create a table of variables.**

   - The torque from F_2 (T_{F2}) = Unknown, to be solved
   - $F_2 = 100$ N
   - $MA_{F2} = 0.2$ m

2. **Select the equation $T = F \cdot MA$, fill in the known values, and solve.**

   $$T_{F2} = F_2 \cdot MA_{F2} = (100 \text{ N}) \cdot (0.2 \text{ m}) = 20 \text{ Nm}$$

Identify the direction of the torque:

> From Figure 8-2, F_2 will cause the body to rotate in the counterclockwise, or +, direction. The torque created by F_2 around the axis is equal to +20 Nm.

Calculate the magnitude of the torque created by $F_3 = 100$ N at the axis indicated:

1. **Create a table of variables.**

   - The torque from F_3 (T_{F3}) = Unknown, to be solved
   - $F_3 = 100$ N
   - $MA_{F3} = 0.5$ m

2. **Select the equation $T = F \cdot MA$, fill in the known values, and solve.**

   $$T_{F3} = F_3 \cdot MA_{F3} = (100 \text{ N}) \cdot (0.5 \text{ m}) = 50 \text{ Nm}$$

Identify the direction of the torque:

> From Figure 8-2, F_3 will cause the body to rotate in the counterclockwise, or +, direction. The torque created by F_1 around the axis is equal to +50 Nm.

The link segment model of the human body is a simplification of the extremely complex anatomy of the muscles and skeleton. The model is useful to describe how muscles act to control the rotation of body segments. Essentially, the model draws heavily from the artistic ability of toddlers to create a stick figure of a person, as in Figure 8-3. A link segment model can also be created for an extremity like the arm or leg, or for an individual segment, like the forearm.

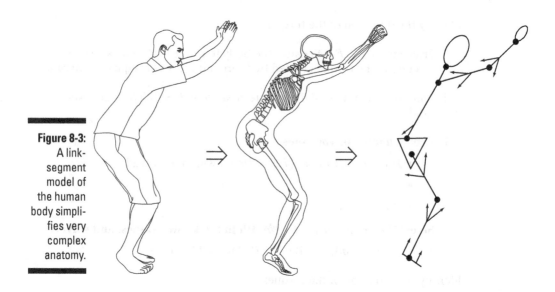

Figure 8-3:
A link-segment model of the human body simplifies very complex anatomy.

The complex anatomy of the bones and muscles of individual segments is simplified by representing each segment as a rigid link, or bar. Separate segments, like the forearm and hand, can be combined depending on the focus of the analysis. The bars representing individual segments meet at joints. Joints are considered to be hinges (like the hinge on a door), and the fixed axis of rotation is the axis of the rotation between adjacent segments.

Muscles cross the joints at some distance from the joints. Although many muscles cross on the front and the back of each joint, the rigid link model simplifies the muscle anatomy so that all the forces produced by all the muscles on one side of the joint is represented with a single force. The single force is called F_M, for muscle force. The F_M is always a pulling force on a segment, shown by vectors in Figure 8-3. Muscles cannot push. (I give you more detail on muscle force in Chapter 15.)

The link segment model shows how the simplified force from all the muscles crossing on one side of a joint creates torque at the joint. The contribution of muscle groups and individual muscles to the torque at joints has been the focus of intense research for many years. Detailed anatomical models of the muscles and segments are created, and the activity of an individual muscle is measured using electromyography (EMG). I describe some of this research in Chapter 17.

Measuring Torque

In the next sections, I use a link segment model consisting of the upper arm and a combined forearm and hand segment to demonstrate torque created at the elbow joint by the F_M of the muscles on the front (anterior surface) of the elbow joint. Then I show the torques created by the weight of the forearm/hand segment and an object held in the hand. I begin by looking at muscle as a torque generator.

Muscling into torque: How muscles serve as torque generators

Figure 8-4 is a link segment model of the arm, consisting of the upper arm and a combined forearm and hand segment. (***Note:*** Because the focus is on the elbow joint, in this model the wrist joint between the forearm and hand is fixed.) Muscle force is *always* a pulling force. The simplified muscle attaches to the forearm 0.05 m from the elbow in this model. The muscle attachment is fixed, and the attachment site does not move as the segments rotate (that would require surgery). The F_M is not a fixed magnitude (I explain how muscle force varies in Chapter 15), but for this example F_M will be set at a constant value of 500 N. I'll show the steps to calculate the torque created by F_M in the three elbow joint positions shown in Figure 8-4.

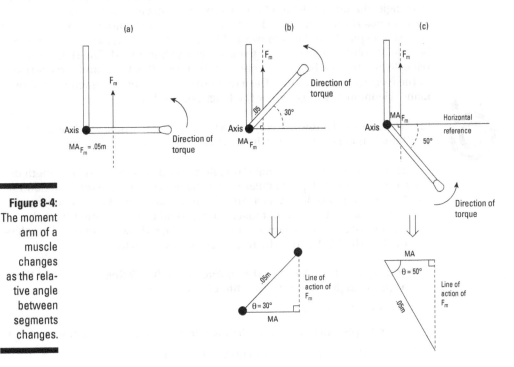

Figure 8-4: The moment arm of a muscle changes as the relative angle between segments changes.

In Figure 8-4a, the relative angle at the elbow is 90 degrees, and the forearm-hand segment is horizontal. Because the muscle inserts at 90 degrees to the forearm in this position, the moment arm of F_M is 0.05 m, a length equal to the distance from the axis of the elbow joint to the point of attachment of the muscle. In this position, the torque created by F_M is calculated as:

$$T = F \cdot MA$$

Now substituting in the values:

$$T_{FM} = F_M \cdot MA_{FM} = (500 \text{ N}) \cdot (0.05 \text{ m}) = 25 \text{ Nm of torque}$$

With the elbow at 90 degrees of flexion, the 500 N of muscle force produces a torque of 25 Nm at the joint. The torque tends to rotate the forearm in the counterclockwise, or +, direction.

With torques around joints, it's best to interpret the torque anatomically. The pulling force from the muscles on the anterior side of the elbow joint creates a torque tending to flex, or decrease the angle at the joint. The torque at the elbow from F_M is described as a flexor torque of 25 Nm.

Because a muscle is firmly attached to segments on both sides of a joint it crosses, the angle of insertion of a muscle on a segment changes as the relative angle changes at the joint crossed by the muscle. The change in the angle of insertion of the muscle alters the length of the moment arm for F_M, as shown in Figure 8-4b. The elbow joint is at 120 degrees of flexion. The forearm is at an absolute angle of 30 degrees above the horizontal. The right-angle triangle created by the line of action of the force, the MA_{FM} and the portion of the forearm segment from the axis to the site of muscle attachment allows using trigonometry to solve for the length of the MA_{FM}

In each example, I show the steps for choosing the trig function to apply. For additional guidance on using trigonometry, see Chapter 2.

I've zoomed in on the right-angle triangle created to show how the length of the moment arm for F_M is calculated. The angle between the forearm-hand and the horizontal is 30 degrees, and this is the angle θ. The distance from the elbow axis to the point of muscle attachment is 0.05 m, and this is the hypotenuse of the right triangle. The MA_{FM} is the side adjacent to θ. To calculate the length of MA_{FM} use the trig function cosine as follows:

1. **Recognize the *MA* as the side *adjacent* to the 30-degree angle of the right triangle and create a table of variables:**

    - θ = 30 degrees

    - Hypotenuse = 0.05 m (distance from axis to the attachment of F_M)

    - Adjacent (MA_{FM}) = Unknown, in meters

2. **Using the process of elimination, select the cosine from SOH CAH TOA.**

$$\cos\theta = \frac{\text{adjacent}}{\text{hypotenuse}}$$

3. **Isolate for the unknown side, the adjacent side, and solve.**

Adjacent = $\cos\theta$ · Hypotenuse = $\cos 30°$ · 0.05 = 0.866 · 0.05 m = 0.04 m

The moment arm for F_M is 0.04 m when the elbow joint is at 120 degrees of flexion. In this position, the torque created by F_M is calculated using $T = F · MA$ and substituting in the known values $T_{FM} = F_M · MA_{FM} = $ (500 N) · (0.04 m) = 20 Nm of torque.

With the elbow joint at 120 degrees of flexion, the 500 N of F_M produces a torque of 20 Nm at the elbow joint tending to rotate the forearm in the counterclockwise, or +, direction, tending to decrease the angle at the elbow joint. The torque at the elbow is described as a flexor torque of 20 Nm.

The same process is used at any joint angle. In Figure 8-4c, the elbow joint is at 40 degrees of flexion. The forearm is at an absolute angle of 50 degrees below the horizontal. A right triangle is created by the line of action of the force, the MA_{FM} and the portion of the forearm segment from the axis to the site of muscle attachment. A zoom of the right triangle clarifies the angle θ between the forearm hand and the horizontal is 50 degrees. The hypotenuse is still 0.05 m, the distance from the elbow axis to the point of muscle attachment. The MA_{FM} is the side adjacent to θ, and the calculation is as follows:

1. **To use the trig function cosine, recognize the MA_{FM} as the side *adjacent* to the 30-degree angle of the right triangle and create a table of variables.**

   - θ = 50 degrees

   - Hypotenuse = 0.05 m (distance from axis to the attachment of *FM*)

   - Adjacent (MA_{FM}) = Unknown, in meters

2. **Using the process of elimination, select the cosine from SOH CAH TOA.**

$$\cos\theta = \frac{\text{adjacent}}{\text{hypotenuse}}$$

3. **Isolate for the unknown side, the adjacent side, and solve.**

Adjacent = $\cos\theta$ · Hypotenuse = $\cos 50°$ · 0.05 = 0.643 · 0.05 m = 0.03 m

The moment arm for F_M is 0.03 m when the elbow joint is at 40 degrees of flexion. In this position, the torque created by F_M is calculated as $T = F \cdot MA$ and substituting in the values $T_{FM} = F_M \cdot MA_{FM} = (500 \text{ N}) \cdot (0.03 \text{ m}) = 15$ Nm of torque.

With the elbow joint at 40 degrees of flexion, the 500 N of F_M produces a torque of 15 Nm at the elbow joint axis. The torque tends flex the joint. The torque at the elbow is described as a flexor torque of 15 Nm.

The moment arm of a muscle changes as the relative angle between segments changes at a joint. Because muscle has a moment arm to a joint axis of rotation, muscle force produces torque at a joint. Muscle serves as a torque generator to affect body movement.

Resisting torque: External torques on the body

Any force applied eccentric to an axis creates a torque. I'll show how torques around the elbow joint are created by the weight of the forearm-hand segment and by the weight of a dumbbell held in the hand. I call the weight of a segment the segment weight (*SW*) and the weight of something held in the hand the handheld weight (*HHW*). Pretty descriptive names, huh?

Figure 8-5 is a link-segment model of the arm like the one I used above to explain muscle torque, but this model shows the *SW* and *HHW* instead of F_M. The forearm-hand segment weight (*SW*) is 18 N, acting downward on the segment with a point of application 0.27 m from the elbow joint. The weight of the transparent dumbbell (*HHW*) is 100 N, and it acts downward on the segment with a point of application 0.36 m from the elbow joint. I'll show the steps to calculate the torques created by the *SW* and the *HHW*.

Figure 8-5: Segment weight and a handheld weight create "loading" torques at the elbow.

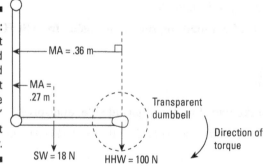

In Figure 8-5, the relative angle at the elbow is 90 degrees, and the forearm-hand segment is horizontal. The *MA* of the *SW* (MA_{SW}) is 0.27 m, a length equal to the distance from the elbow axis to the point of attachment of the *SW*. The torque created by *SW* (T_{SW}) is calculated as follows:

$$T = F \cdot MA$$

Now substitute in the values:

$$T_{SW} = SW \cdot MA_{SW} = (-18 \text{ N}) \cdot (0.27 \text{ m}) = -4.9 \text{ Nm of torque}$$

With the elbow at 90 degrees of flexion, the SW of –18 N produces a torque of –4.9 Nm at the elbow. The torque tends to cause rotation of the forearm-hand segment in the clockwise, or –, direction, increasing the angle of the elbow joint (extension). The *SW* creates an extensor torque at the elbow. The torque at the elbow produced by the *SW* is described as an extensor torque of 4.9 Nm.

The torque at the elbow joint created by the *HHW* is calculated using the same steps. The *MA* of the *HHW* (MA_{HHW}) is 0.36 m, and the *HHW* = –100 N. Torque is calculated as $T = F \cdot MA$, so $T_{HHW} = HHW \cdot MA_{HHW} = (-100 \text{ N}) \cdot (0.36 \text{ m}) = -36 \text{ Nm}$. The sign of the torque corresponds with the tendency of the *HHW* to rotate the forearm in the clockwise (–) direction, an extensor torque. The *HHW* creates an extensor torque of 36 Nm.

The net torque (ΣT_{elbow}) at the elbow joint is calculated by summing the torques at the elbow joint. $\Sigma T_{\text{elbow}} = T_{SW} + T_{HHW} = (-4.9 \text{ Nm}) + (-36 \text{ Nm}) = -40.9 \text{ Nm}$. The weight of the forearm-hand segment and the handheld weight create an extensor torque at the elbow of 40.9 Nm.

Setting the origin of the coordinate system used to measure the moment arms at the axis of the joint will make things work out mathematically. The *MA* is a negative value if the *MA* is to the left of the joint axis (*–MA*), and the *MA* is a positive value if the *MA* is to the right of the joint axis (*+MA*).

Expanding on Equilibrium: Balanced Forces and Torques

Equilibrium means to be in balance. In mechanics, a body is in equilibrium when there are no unbalanced forces or unbalanced torques acting on the body. The sum of the forces in all directions is zero, and the sum of all the torques around all axes is zero. Mathematically, $\Sigma F_{\text{Direction}} = 0$ and $\Sigma T_{\text{Axis}} = 0$.

For equilibrium, the $\Sigma F = 0$ in all directions (vertical, antero-posterior, and medial lateral) and $\Sigma T = 0$ around all possible axes. I'm only looking at force in one direction (vertical) and torque around one axis of the link-segment model.

Consider the forearm-hand holding the transparent barbell in Figure 8-6. I've simplified the link-segment model by removing the upper-arm segment. The forearm is at rest in the horizontal position. The forces on the segment are the F_M, the SW, and the HHW. I've added another external vertical force at the elbow joint to represent the force from contact between the forearm-hand segment and the upper arm. I call this force JF_v, for joint force in the vertical direction. JF_v is centric to the elbow axis, so it doesn't create a torque at the elbow. The SW is –18 N, with an MA of 0.27 m. The HHW is –100 N with an MA of 0.36 m. The MA for the $F_M = 0.05$ m. Because the forearm-hand segment is in equilibrium, I'll use $\Sigma T_{elbow} = 0$ to calculate the magnitude of the F_M and then use $\Sigma F_{vertical} = 0$ to solve for JF_v.

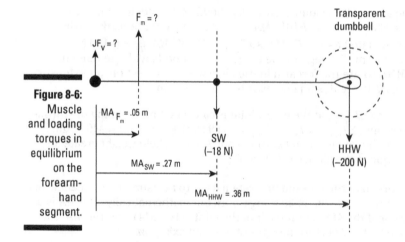

Figure 8-6: Muscle and loading torques in equilibrium on the forearm-hand segment.

Figure 8-6 is a free-body diagram (FBD) of the forearm-hand segment, showing all the external forces acting on the body. Creating an FBD is explained in more detail in Chapter 4.

Create a table of variables:

F_M = Unknown (the force of the muscle), to be solved

$MA_{MF} = 0.05$ m (the MA for the muscle force)

$SW = -18$ N (the weight of the segment)

$MA_{SW} = 0.27$ m (MA for the SW)

$HHW = -100$ N (the weight of the dumbbell held in the hand)

$MA_{HHW} = 0.36$ m (the MA for the HHW)

JF_v = Unknown (the joint force between the forearm-hand and the upper arm), to be solved

Knowing that $\Sigma T_{elbow} = 0$, identify all the torques at the elbow joint from eccentric forces (F_M, HHW, and SW) and expand the equation:

$$\Sigma T_{elbow} = 0 = T_{FM} + T_{HHW} + T_{SW}$$

Now isolate for the T_{FM}:

$$-T_{FM} = T_{HHW} + T_{SW}$$

Now expand each of the torques to $F \cdot MA$:

$$-(F_M \cdot MA_{MF}) = (HHW \cdot MA_{HHW}) + (SW \cdot MA_{SW})$$

Now isolate for F_M, multiply both sides by -1, fill in the known values, and solve:

$$F_M = -\left(\frac{(HHW \cdot MA_{HHW}) + (SW \cdot MA_{SW})}{MA_M} \right)$$

$$= -\left(\frac{([-100 \text{ N}] \cdot [0.36 \text{ m}]) + ([-18 \text{ N}] \cdot [0.27 \text{ m}])}{0.05 \text{ m}} \right)$$

$$= -\left(\frac{(-36 \text{ Nm}) + (-4.9 \text{ Nm})}{0.05 \text{ m}} \right) = \left(\frac{-40.9 \text{ Nm}}{0.05 \text{ m}} \right) = +818 \text{ N}$$

To maintain the forearm in equilibrium, the F_M must be $+818$ N. Note that F_M ($+818$) is greater in magnitude than the sum of the SW (-18 N) and the HHW (-100 N). The muscles of the body have relatively short moment arms compared to other loads applied to the segment. To produce the flexor torque large enough to offset the extensor torques created by the SW and the HHW, the magnitude of the F_M must be greater than the sum of the forces.

Muscles in the body are said to be at a mechanical disadvantage because they have relatively short moment arms compared to the loading forces applied to the body.

With F_M solved, I can now solve for JF_v using $\Sigma F_{vertical} = 0$. I'll begin by revising the table of variables from above, adding in the calculated value of F_M and getting rid of the information on the moment arms:

$F_M = +818$ N (solved above)

$SW = -18$ N

$HHW = -100$ N

$JF_V =$ Unknown, to be solved

Knowing that $\Sigma F_{vertical} = 0$, expand the equation, isolate for the unknown value, fill in the known values, and solve:

$$\Sigma F_{vertical} = 0 = F_M + SW + HHW + JF_v$$
$$JF_V = -F_M - SW - HHW = -(+818\ N) - (-18\ N) - (-100\ N) = -700\ N$$

The $JF_v = -700$ N shows the upper arm is pushing *down* on the forearm-hand segment at the elbow joint. The upper arm must push down on the forearm-hand segment to offset the large muscle force pulling upward. With no unbalanced force acting on the forearm-hand segment, the segment will remain in the horizontal position.

Locating the Center of Gravity of a Body

The center of gravity is defined as the point of application of a body's weight. It's sometimes called the center of mass, and the two terms are interchangeable. The location of the center of gravity is easily identified for a body made of one material, like a board of wood — the center of gravity is typically at the middle of the physical body. However, because the human body consists of a lot of different materials, unequally distributed throughout the rather odd-shaped segments of the body, the center of gravity isn't at the middle of the human body. At rest in the anatomic position, the center of gravity of the human body is located just below the navel, or belly button. This location is about 55 percent of the person's height above the ground. The concept of equilibrium was used to identify this location. I explain how, using Figure 8-7.

Figure 8-7:
Using equilibrium to identify the location of the center of gravity.

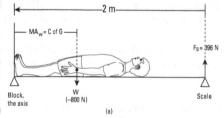
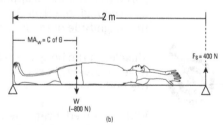

I said "about 55 percent" in describing the center of gravity location because the actual location varies among people — 55 percent is an average value. For females, the center of gravity tends to be a little lower (about 54 percent of body height above the ground), while for males, the center of gravity is a little higher (about 56 percent of body height above the ground). The precise location differs among individual females and individual males, depending on how an individual's mass is distributed. A good approximation of the center of gravity location in the anatomic position is "just below the navel, about 55 percent of body height above the ground."

Figure 8-7a shows a person lying on a board 2 m long. The board rests on a pointed block at one end and on a scale with a sharp point at the other end. The person is 1.8 m tall and weighs 800 N, and lies on the board with the bottoms of both feet aligned with the sharp block and both arms at the side. The scale has a reading of +396 N, the magnitude of the force of the scale pushing up on the end of the board. The block at the person's feet is set as the axis of rotation for the calculation of the torques created by the eccentric forces of the person's weight (W) and the force of the scale (F_s). A clockwise (−) torque around the axis is created by the person's body weight pulling downward, and a counterclockwise (+) torque is created around the axis by the force of the scale pushing upward. The person and the board are at rest — the person/board "body" is in equilibrium, so the $\Sigma T_{axis} = T_{Fs} + T_W = 0$. The center of gravity is the point of application of the body weight. The location of the center of gravity is solved as follows:

Create a table of variables:

> F_s = +396 N (the force reading from the scale)
>
> MA_s = +2 m (the distance of the point of the scale from the axis of rotation)
>
> W = −800 N (the weight of the person, weight is always directed down)
>
> MA_W = The location of the center of gravity; unknown, to be solved

Knowing that ΣT_{axis} = 0, identify all the torques at the axis from eccentric forces and expand the equation:

> $\Sigma T_{axis} = 0 = T_{Fs} + T_W$

Now isolate for the T_W:

> $-T_W = T_{Fs}$

Now expand each of the torques to $F \cdot MA$:

> $-(W \cdot MA_W) = (F_s \cdot MA_{Fs})$

Now isolate for the MA_W, multiply both sides by -1, fill in the known values, and solve:

$$MA_W = -\left(\frac{(F_s \cdot MA_s)}{W}\right) = -\left(\frac{(+396 \text{ N}) \cdot (+2 \text{ m})}{-800 \text{ N}}\right)$$

$$= -\left(\frac{+792 \text{ Nm}}{-800 \text{ N}}\right) = -(-0.99 \text{ m}) = +0.99 \text{ m}$$

The center of gravity of the person is located 0.99 m above the feet. Calculating the ratio of this location to the person's height (1.8 m) as 0.99 m ÷ 1.8 m = 0.55 shows the center of gravity is located at 55 percent of the person's height above the feet. The center of gravity is sometimes referred to as "the balance point of the body" because the torques created by body mass sum to zero when calculated around the center of gravity.

If a segment or segments are shifted, body mass is shifted, causing the center of gravity of the body to move in the direction of the new position of the segments. In Figure 8-7b, the person lies on the board with both arms raised above the head. The mass of the arms is now farther from the feet. The center of gravity of the body shifts upward, away from the feet. The scale reading is now 400 N. To calculate the change in position of the center of gravity of the body, I'll solve again for the $\Sigma T_{axis} = 0$.

Create a table of variables:

$F_s = +400$ N (the new force reading from the scale)

$MA_s = +2$ m (the distance of the point of the scale from the axis of rotation)

$W = -800$ N (the weight of the person, unchanged although segments shifted)

$MA_W =$ The location of the center of gravity; unknown, to be solved

Re-solve for $\Sigma T_{axis} = 0$, using the new information:

$$\Sigma T_{axis} = 0 = T_{Fs} + T_W$$

Now isolate for the T_W:

$$-T_W = T_{Fs}$$

Now expand each of the torques to $F \cdot MA$:

$$-(W \cdot MA_W) = (F_s \cdot MA_{Fs})$$

Now isolate for the MA_W, multiply both sides by –1, fill in the known values, and solve:

$$MA_W = -\left(\frac{(F_s \cdot MA_s)}{W} \right) = -\left(\frac{(+400 \text{ N}) \cdot (+2 \text{ m})}{-800 \text{ N}} \right)$$
$$= \left(\frac{+800 \text{ Nm}}{-800 \text{ N}} \right) = -(-1.00 \text{ m}) = +1.00 \text{ m}$$

With the arms overhead, the center of gravity of the person is now located 1.0 m above the feet. The center of gravity moved upward 0.01 m (1 cm) in the body. Calculating the ratio of this location to the person's height (1.8 m) as $1.00 \div 1.8$ m = 0.56 shows the center of gravity is now located at about 56 percent of the person's height above the feet. The arms were raised above the head, farther from the feet, and the center of gravity moved upward. The center of gravity of a body is not a fixed point — it reflects the posture of the body.

Each individual segment has its own center of gravity. These were originally identified in studies conducted for the U.S. Air Force in the 1950s. Cadavers were carefully dissected, separating the segments at joints. The individual segments were hung on hooks to locate the "balance point" of each segment; when the hook was inserted and the segment no longer rotated (equilibrium was attained), the center of gravity was identified. The location of the center of gravity was identified as a percentage of the segment length from the proximal and the distal end of the segment, and tables of these values were created and published. This experiment, although macabre sounding, provided data that are used regularly in biomechanics.

Chapter 9

Angling into Rotation: Angular Kinematics

*A*ngular kinematics is the branch of mechanics concerned with the description of a rotating body, which includes swinging, spinning, and twisting. The focus of the analysis is a body, and the body can be an implement (such as a baseball bat), the entire human body (such as a gymnast completing a somersault), an individual segment (such as the motion of the hand during a golf swing), or two segments moving relative to each other (such as the motion at the knee between the thigh and the shank when running).

This chapter explains the terms used in angular kinematics to describe how far, how fast, and how consistently (speeding up or slowing down) the body rotates. Because all human motion is caused by the coordinated rotation of individual segments, I also describe the important relationship between angular motion and linear motion.

Measuring Angular Position

The left side of Figure 9-1 shows that an angle is created when two lines intersect. The angle created is represented by the Greek letter θ (theta). But an angle can also be created between a single line and a reference plane, as in the right side of Figure 9-1.

Figure 9-1:
Angles
created
between
two lines
(left) and a
line and a
plane (right).

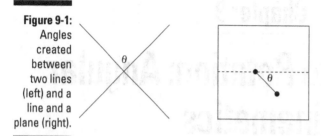

Two units can be used to describe a measured angle. The most familiar is the degree (°). There are 360 degrees in a circle. A 90-degree angle is called a *right angle*. When two line segments join to create a straight line, the angle is 180 degrees. Degrees are used most commonly by everyday practitioners of biomechanics such as sport coaches who might cue a player to flex his knees to 90 degrees, and athletic trainers who might measure the amount of knee flexion a patient has after an operation.

A less common, but equally important, unit used for angular measurement is the *radian*. Look at the circle in Figure 9-2. The line from the circle's center to its circumference is the *radius* (abbreviated *r*). If a piece of string equal to the length of the radius is laid along the circumference, an angle is created when a radius is drawn between each end of the arc. This angle is called a *radian* (*rad* for short). Because the length of a circle's circumference is defined as $2\pi r$, and there are 360 degrees in a circle, this angle is always 57.3 degrees:

$$2\pi r = 360°$$

$$1 \text{ radian} = \frac{360°}{2\pi} = \frac{360°}{2 \times 3.14} = \frac{360°}{6.28} = 57.3°$$

Figure 9-2:
A *radian* is
the angle
created
when the
radius of a
circle is laid
along the
circumfer-
ence of the
circle.

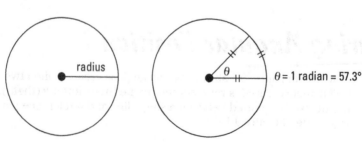

The radian is important because it links a linear measure of a circle, the radius, to the angular measure of a circle. Angles must be measured in, or converted to, radians when looking at the relationship between angular and linear motion, as presented later in this chapter.

Radians yield smaller numbers for a measured angle, but they aren't often used by everyday practitioners of biomechanics. Radians are used to measure angles when relating angular and linear motion.

Divide an angle measured in degrees by 57.3 to convert the angular measurement to radians.

Two types of angles are of interest in biomechanics: relative angles and absolute angles. Figure 9-3 shows both relative angles and absolute angles as measured on a runner.

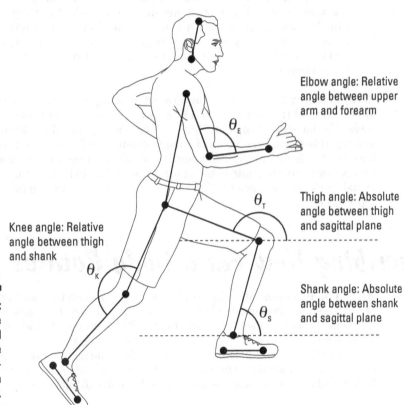

Elbow angle: Relative angle between upper arm and forearm

θ_E

Thigh angle: Absolute angle between thigh and sagittal plane

θ_T

Knee angle: Relative angle between thigh and shank

θ_K

Shank angle: Absolute angle between shank and sagittal plane

θ_S

Figure 9-3: Relative angles and absolute angles measured on a runner.

✓ **Relative angles:** A *relative angle* is created when two lines meet or intersect. *Relative* means the angle is being measured from one line to the other line. In biomechanics, the most common relative angles are those created when two body segments meet at a joint, such as the thigh and *shank* (lower leg) meeting at the knee joint, or the forearm and upper arm meeting at the elbow joint. The relative angle of the knee joint and the elbow joint are shown on Figure 9-3. There are no "lines" on the segments, so they're drawn on or visualized. All the joint centers are identified (represented with the dots), and a line is drawn between the joint centers to identify the segment. A joint angle is measured as the relative angle between adjacent segments.

✓ **Absolute angles:** An *absolute angle* is created when a single line is measured against a fixed external reference, often one of the defined reference planes (see Chapter 2). The absolute angle of any segment can be measured from a horizontal or vertical line drawn on the reference plane. Figure 9-3 shows the absolute angles of the thigh and the shank (lower leg) of a runner measured against the sagittal plane. Absolute angles are used when the orientation of an individual segment is of interest, such as foot angle at contact with a ball when kicking, trunk lean during running, or upper arm orientation when throwing.

Always identify how the angle of interest is being measured, because the intersection of the two lines or a line and a plane can be measured two ways. The knee joint angle in Figure 9-3 can be measured as (1) the angle measured between the backs of each segment or (2) the angle measured between the fronts of each segment. Obviously, recognizing how a relative or absolute angle is being measured is important so that you can visualize the correct orientation of the segment(s) when you read the reported angle.

Describing How Far a Body Rotates

When a body rotates, all points on the body go through the same amount of degree (or radian) change in angular position during the rotation. We use the same method of measuring an angular change in position whether we're analyzing the rotation of a relative angle, an absolute angle, or an entire body (such as when a diver performs a somersault in the air). In this section, I explain how to calculate and describe "how far" a body rotates, including both angular distance and the more useful angular displacement.

Angular distance

Angular distance is equal to the entire length of the angle traveled between the first and the final angular positions. This includes the entire path of rotation of the body. Consider the change in the relative angle of the knee joint shown in Figure 9-4. At the start position, indicated with θ_1, the knee is straight, measured as an angle of 180 degrees. The knee is then bent (flexed) to position θ_2, an angle of 90 degrees, and finally straightened (extended) to position θ_3, an angle of 135 degrees. The angular distance is the sum of the length of the angle traveled from θ_1 to θ_2 (90 degrees) and the length of the path traveled from θ_2 to θ_3 (45 degrees), an angular distance of 135 degrees.

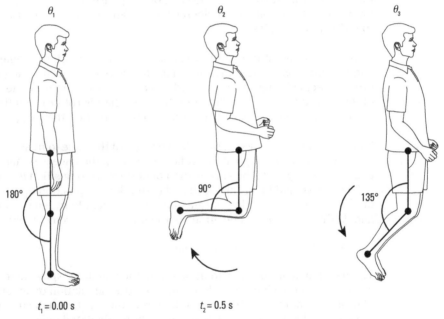

Figure 9-4: The knee is rotated from position θ_1 to position θ_2 and then to position θ_3.

θ_1 — 180° — $t_1 = 0.00$ s

θ_2 — 90° — $t_2 = 0.5$ s

θ_3 — 135°

Although this made for a good example, angular distance is not frequently used to describe the rotation of segments or joints. *Angular displacement*, covered in the next section, is a more appropriate descriptor. However, *angular distance* is commonly used to describe the full- or partial-body rotations that occur in sports like diving and gymnastics. When a diver leaves the board upright and completes a forward somersault to enter the water headfirst, the dive is described as a one and a half, with *one and a half* referring to the number of revolutions completed. The *one* is the full revolution for the diver to rotate in the air to the head up position, and the *half* is the incomplete revolution for the diver to rotate to the head down position to enter the water.

Angular displacement

Angular displacement is the angle formed between the initial position of the body and the final position of the body. That is, angular displacement is the net change in position of an angle. Angular displacement is represented with $\Delta\theta$, with Δ (delta) the symbol for "change in." $\Delta\theta$ is calculated as $\theta_f - \theta_i$, where θ_f is the *final angle* (the angle at the end of the phase being analyzed) and θ_i is the *initial angle* (the angle at the start of the phase being analyzed). The units for angular displacement are degrees or radians, depending on the units used to measure the angles.

$\Delta\theta$ is a vector quantity, so both the *magnitude* (how much the angle changes) and the *direction* (which way the angle changes) are reported. Angular direction is specified relative to the hands on a clock. By definition, the negative direction is *clockwise* (the direction the clock hands rotate), and the positive direction is *counterclockwise* (opposite to the direction the hands rotate).

Look again at the change in the relative angle of the knee joint as shown in Figure 9-4. At the start position, indicated with θ_1, the knee is straight, a knee angle of 180 degrees. The knee is then bent (flexed) to position θ_2, a knee angle of 90 degrees, and finally straightened (extended) to position θ_3, a knee angle of 135 degrees. Calculating the angular displacement between positions θ_1 and θ_3, the start and end positions of the total knee movement, gives

$$\Delta\theta = \theta_f - \theta_i = \theta_3 - \theta_1 = 135° - 180° = -45°$$

The net change in the knee joint position, the angular displacement, was 45 degrees in the clockwise direction. Note the calculation of angular displacement ignores any changes in the direction of rotation occurring between the specified initial and final positions. When calculated over a total movement like the knee motion above, angular displacement covers up useful information. The knee moved in two different directions during the total movement, but the calculation of $\Delta\theta$ suggests the knee moved only a small portion of the complete movement, and that rotation occurred in only one direction. Calculating angular displacement is most useful when the change in angle during a smaller portion of the movement is considered, typically choosing a portion of the movement when no change in direction occurs.

Calculating angular displacement between θ_1 (180°) and θ_2 (90°) gives:

$$\Delta\theta = \theta_f - \theta_i = \theta_2 - \theta_1 = 90° - 180° = -90°$$

The angular displacement was 90 degrees in the clockwise (–) direction. The direction of joint motion was the same during the entire period analyzed, and no part of the motion during the period was ignored. With the view of the performer in Figure 9-4, the – sign can be interpreted as bending, or *flexion*, of the knee joint.

Calculating angular displacement between θ_2 (90 degrees) and θ_3 (135 degrees) gives

$$\Delta\theta = \theta_f - \theta_i = \theta_3 - \theta_2 = 135° - 90° = +45°$$

The angular displacement was 45 degrees in the counterclockwise (+) direction. With the view of the performer in Figure 9-4, the + sign can be interpreted as the straightening, or *extension*, of the knee joint.

The calculated angular displacement indicates not only how far the joint rotated (magnitude) but also what type of joint action occurred (direction). But obviously your position when viewing the movement affects what is perceived as clockwise and counterclockwise. (Check this by redoing the calculations as if you were looking at the person when standing on the opposite side — although the magnitude of the displacements will be the same, the directions will be opposite and will lead to the same anatomical interpretation of the action.)

Describing How Fast a Body Rotates

How fast a body rotates is a critical factor in performance. Generally, better performance is characterized by faster rotations. Fast actions at the ankle, knee and hip joints affect the size of the force applied to the ground during gait. Faster joint actions during a throw or a swing affect how fast a ball is thrown or a baseball bat is swung. Faster spinning or twisting determines if a diver or a gymnast will complete a required trick. In this section, I explain the calculation of how fast a body rotates.

Angular speed

Angular speed is the angular distance covered during the rotation divided by the time taken to move through the angular distance. Angular speed is a scalar quantity and describes only how fast a body rotates. Angular speed is not frequently calculated in biomechanics, because specifying the direction of the angular motion is usually as important as specifying how fast rotation occurs when analyzing performance.

The term *angular speed,* or simply *speed,* is used to describe the rate of rotation when the direction of the rotation is known. In these cases, speed actually refers to the magnitude of the angular velocity, or how fast a body is rotating. Angular velocity is defined and described in the next section.

Angular velocity

Angular velocity is a vector quantity describing both how fast and in what direction a body is rotating. The symbol for angular velocity is the Greek letter ω (omega). To calculate angular velocity, the angular displacement ($\Delta\theta$) of a body is divided by the time it took for the rotational displacement to occur (Δt), or as an equation $\omega = \dfrac{\Delta\theta}{\Delta t}$.

The units of angular velocity are the units of angular displacement (degrees or radians or revolutions) divided by the units of time (seconds) or degrees per second (°/s) or radians per second (rad/s) or revolutions per second (rev/s).

$\Delta\theta$ is calculated as $\theta_f - \theta_i$, the net change in position during the period being studied. Δt is the duration of the period. Δt is calculated as $t_f - t_i$, (read "time final minus time initial"), the time that passed between the start and the end of the period.

In Figure 9-4, t_1, t_2, and t_3 indicate the time when each angular position was identified, as if a stopwatch were used to measure the time while the leg was moved. Calculating ω between t_1 and t_2 goes like this:

1. **Identify the known values and create a table of variables:**

   - $t_f = t_2 = 0.5$ seconds (s)
   - $t_i = t_1 = 0.0$ s
   - $\theta_f = \theta_2 = 90$ degrees
   - $\theta_i = \theta_1 = 180$ degrees

2. **Select the equation and substitute the given values.**

$$\omega = \frac{\theta_f - \theta_i}{t_f - t_i} = \frac{90° - 180°}{0.5\text{ s} - 0.0\text{ s}} = \frac{-90°}{0.5\text{ s}} = -180° / \text{s}$$

The angular velocity during the period is 180 degrees/second in the clockwise (–) direction. With the view of the performer in Figure 9-4, the – sign can be interpreted as the bending, or *flexion,* of the knee joint. Report a flexion angular velocity of 180 degrees/second.

Calculating ω between t_2 and t_3 goes like this:

1. **Identify the known values and create a table of variables:**
   - $t_f = t_3 = 0.75$ seconds (s)
   - $t_i = t_2 = 0.5$ s
   - $\theta_f = \theta_3 = 135$ degrees
   - $\theta_i = \theta_2 = 90$ degrees

2. **Select the equation and substitute the given values.**

$$\omega = \frac{\theta_f - \theta_i}{t_f - t_i} = \frac{135° - 90°}{0.75 \text{ s} - 0.5 \text{ s}} = \frac{+45°}{0.25 \text{ s}} = +180° / \text{s}$$

The angular velocity during the period is 180 degrees/second in the counterclockwise (+) direction. With the view of the performer in Figure 9-4, the + sign can be interpreted as the straightening, or *extension,* of the knee joint. Report an extension angular velocity of 180 degrees/second.

The angular velocity calculated is the *average* velocity of the body over the entire period Δt. The angular velocity may actually vary over that interval. To calculate the instantaneous angular velocity, or how fast the body is rotating at an instant in time, the time period Δt is made very small. In a quantitative analysis of performance (see Chapter 17), when technology is used to record and analyze performance, a Δt of 0.005 seconds is commonly used to calculate the instantaneous velocity throughout the performance.

Speeding Up or Slowing Down: Angular Acceleration

Angular acceleration is the rate of change of angular velocity with respect to time. The symbol for angular acceleration is the Greek letter α (alpha). To calculate angular acceleration, the change in the angular velocity ($\Delta\omega$) of a body is divided by the time it took for the change in angular velocity to occur (Δt), or as an equation $\alpha = \frac{\Delta\omega}{\Delta t}$.

$\Delta\omega$ is calculated as $\omega_f - \omega_i$, the change in angular velocity during the period being studied. Δt is the duration of the period. Δt is calculated as $t_f - t_i$ (read "time final minus time initial"), the time that passed between the start and the end of the period being analyzed.

The units of angular acceleration are the units of angular velocity (°/s or rads/s or rev/s) divided by units of time (seconds) or degrees per second per second (°/s/s) or radians per second per second (rads/s/s) or revolutions per second per second (rev/s/s).

Figure 9-5 shows a person raising his arm above his head (a) and then lowering his arm back to his side (b). In both a and b, ω is 0°/s at the top and bottom of the motion. At the midpoint of a, the raising phase, the ω of the arm is +180°/s. At the midpoint of b, the lowering phase, the ω of the arm is –180°/s. t_1 through t_6 indicate the time at which each ω was measured.

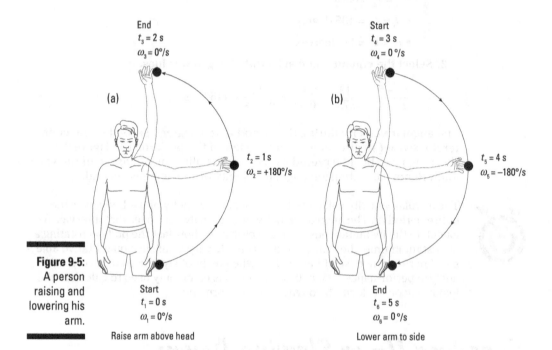

Figure 9-5:
A person raising and lowering his arm.

Angular acceleration can be calculated over two phases as the arm rises: as ω increases (speeds up) from 0 to +180°/s, and then as ω decreases (slows down) from +180°/s to 0. Similarly, angular acceleration can be calculated over two phases as the arm lowers: as ω increases (speeds up) from 0 to –180°/s, and then as ω decreases (slows down) from –180°/s to 0. These calculations are as follows:

Interval	ω_i	t_i	ω_f	t_f	$\Delta\theta\ \omega = \omega_f - \omega_i$	$\Delta t = t_f - t_i$	α
Rise							
t_1 to t_2	0	0	+180	1	180 – 0 = 180	1 – 0 = 1	+180 ÷ 1 = +180
t_2 to t_3	+180	1	0	2	0 – (+180) = –180	2 – 1 = 1	–180 ÷ 1 = –180
Lower							
t_4 to t_5	0	3	–180	4	–180 – 0 = –180	4 – 3 = 1	–180 ÷ 1 = –180
t_5 to t_6	–180	4	0	5	0 – (–180) = +180	5 – 4 = 1	+180 ÷ 1 = +180
Units	°/s	s	°/s	s	°/s	s	°/s/s

Both the raising and lowering phases show periods of + and − acceleration. But the interpretation of the + and − accelerations is opposite in each phase. When raising the arm by rotating in the + direction, $+\alpha$ reflects speeding up from 0 to 180°/s, and $-\alpha$ reflects slowing down from +180 to 0°/s. When lowering the arm by rotating in the − direction, $-\alpha$ reflects speeding up from 0 to −180°/s, and $+\alpha$ reflects slowing down from −180 to 0°/s. Unlike $\Delta\theta$ and ω, the sign of the calculated acceleration does not immediately give information on the motion. The direction of the angular velocity must also be considered. Three rules for interpreting angular acceleration are

✔ When the sign of ω_f and the sign of α are the same, the ω increases (rotation is speeding up).

✔ When the sign of ω_f and the sign of α are the opposite, the ω decreases (rotation is slowing down).

✔ Whenever ω_f is 0, regardless of the sign of α, the ω decreases (rotation is slowing down).

In mechanics, the term *acceleration* means a change in the motion, or velocity, of the body. This can be either speeding up or slowing down. So, *angular acceleration* refers to speeding up or slowing down the angular velocity of a body. Use the sign of the acceleration and the direction of the motion to interpret a calculated acceleration.

Relating Angular Motion to Linear Motion

To throw a ball, coordinated angular motions of body segments are used to increase the speed of the hand holding the ball. At release, the ball becomes a projectile, and its path through the air is described using linear kinematics (see Chapter 5). The relationship between angular motion and linear motion explains how rotations increase linear velocity and lead to success in all throwing and kicking activities (baseball, softball, soccer) and in activities when an implement (club, stick, or bat) is swung to strike a target object (golf ball, puck, or baseball). This section explains the relationships between linear and angular kinematics, beginning with displacement, moving on to velocity, and ending with acceleration.

This section links measures of angular kinematics and linear kinematics. Chapter 5 explains linear speed, linear velocity, and linear acceleration in more detail and also covers the topic of projectile motion.

Angular displacement and linear displacement

The definition of rotation is that all points on the rotating body undergo the same angular displacement. But points on the rotating body travel different linear distances during the same period of rotation, depending on how far a point is from the axis of rotation. The *axis of rotation* is the identified line around which a body rotates. The axis of rotation for a gymnast swinging on the high bar is easily identified because the hands grip the bar. However, in other swinging movements, there is often no clearly observable axis, and the axis must be specified and used for the analysis.

Figure 9-6 shows a golfer at two instances during a swing. The focus of this section is the golfer's arms and club moving as one body during the downswing. The axis of rotation is set between the shoulders — this is not a rigid, visible axis, but the arms and club are defined as rotating around an imaginary axis at this location. Both the hands and the club head undergo 45 degrees (0.76 radians) of angular displacement during the downswing, because they move as a single body (at least with good golfers!).

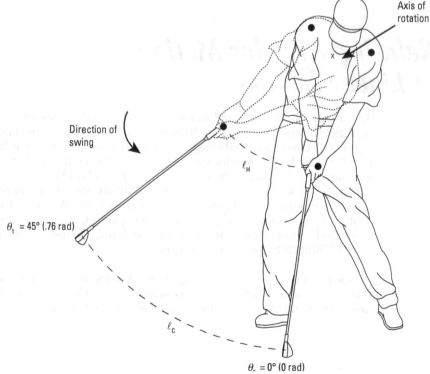

Axis of rotation

Direction of swing

ℓ_H

Figure 9-6: The hands and the club head undergo the same angular displacement but travel different linear distances.

$\theta_1 = 45° (.76\ \text{rad})$

ℓ_c

$\theta_2 = 0° (0\ \text{rad})$

But notice the difference between the linear distance of the hands and the club head, indicated with the dashed lines ℓ_H for the hands and ℓ_C for the club head. The linear distance traveled by the hands is less than the linear distance traveled by the club head.

The linear distance traveled by a point on a body, ℓ, is related to the angular displacement of the rotating body, $\Delta\theta$ and by the length of the radius (r). In this case, r is the distance from the axis of rotation to a specified point on the body. For the golfer, the radius from the axis of rotation to the hands is shorter than the radius from the axis of rotation to the club head.

The equation linking linear distance to angular displacement and the radius of rotation is $\ell = \Delta\theta r$. Because all points on a rotating body undergo the same $\Delta\theta$, the equation reinforces what can be seen in Figure 9-6: A point farther away from the axis of rotation (a point with a longer r) travels farther than a point closer to the axis of rotation (a point with a shorter r).

Angular velocity and linear velocity

The golfer in Figure 9-6 uses the club to strike the ball, intending to drive the ball down the fairway. Implements are used similarly in tennis (a racket), hockey (a stick), and softball (a bat) to play those games. Typically, the success of the golf drive depends on the linear velocity of the club head at the instant of contact between the club head and the ball. In this section, I explain the relationship between the angular velocity of a rotating body and the linear velocity of a point on the body.

The linear distance ℓ traveled by a point on a rotating body is related to the angular displacement $\Delta\theta$ of the body by the equation $\ell = \Delta\theta r$, where r is the radius from the axis of rotation to the point. If both sides are divided by the time it takes for the rotation to occur, Δt, the equation becomes

$$\frac{\ell}{\Delta t} = \frac{\Delta\theta}{\Delta t} r$$

The left side of this equation, $\frac{\ell}{\Delta t}$, is the linear speed (see Chapter 5) of the point. Since $\frac{\Delta\theta}{\Delta t}$ is the equation for the angular velocity of a body (ω), the right side of the equation reduces to ωr. Rewriting the equation in this new form gives

Linear Speed $= \omega r$

For two points on a rotating body, the point farther from the axis of rotation has a higher linear speed. This, of course, makes sense because the point farther from the axis of rotation travels a greater distance in the same period of time.

At an instant in time, the linear speed of the point is the same as the linear velocity of the point. Speed is equal to the magnitude of a velocity vector. The linear velocity of a rotating object is called the tangential velocity (v_T). When throwing a ball, the v_T of the ball at the moment the ball is released from the hand is the initial velocity value that determines the parabolic flight of the projectile, as solved for with projectile motion (see Chapter 5). When hitting a golf ball with a club, the v_T of the club head at the instant of contact with the ball determines the success of the golf shot. v_T at the instant of contact or release is important in all striking and throwing activities.

v_T is called tangential velocity because the direction of this linear velocity is at a tangent to the circular path of the rotating body.

The equation for tangential velocity is $v_T = \omega r$, the same equation as for linear speed. Figure 9-7 shows the arm of a baseball pitcher at two instants when throwing a 70 mph fastball. To throw a 70 mph fastball means the ball leaves the pitcher's hand moving 70 mph; this is called the *moment of release* and is shown in Figure 9-7. Because the ball and the hand move together, the hand is also moving at 70 mph at the moment of release. I'll assume at release that the arm is one rigid segment because the elbow is in a fixed position. The radius from the shoulder joint (the axis of rotation of the arm) to the ball and the hand is 0.75 meters. I'll use the equation for v_T to calculate the angular velocity (ω) of the arm at the shoulder joint at release of the ball.

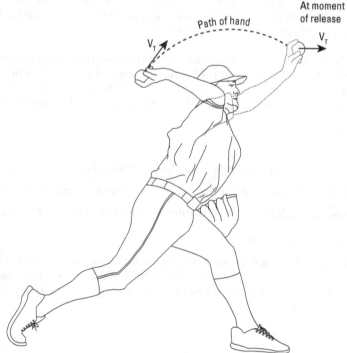

Figure 9-7: A pitcher throwing a 70 mph fastball.

1. **Create a table of variables.**

   - ω = unknown, to be calculated
   - v_T = 70 mph, which is about 34 m/s
   - r = 0.75 m

2. **Select the appropriate equation, isolate for the unknown value, plug in the know values, and solve.**

   $$v_T = \omega r$$

   $$\omega = \frac{v_T}{r} = \frac{34 \text{ m/s}}{0.75 \text{ m}} = 45.3$$

The value of 45.3 is the angular velocity of the arm in radians per second (rad/s). To calculate the angular velocity in degrees per second, multiply ω in rad/s by 57.3, because there are 57.3 degrees in a radian:

$$\omega°/s = \omega \text{ rad/s} \times 57.3° = 45.3 \times 57.3 = 2{,}595.69°/s$$

Angular acceleration and linear acceleration

A point on a rotating body undergoes two types of linear acceleration. One acceleration is related to the magnitude of v_T, and is called *tangential acceleration*. (Isn't mechanics a descriptive language?) The second acceleration is related to the changing direction of v_T as the body rotates and is called *centripetal acceleration*. These two accelerations are explained in this section.

Tangential acceleration

v_T is the linear velocity of a point on a rotating body and is calculated as $v_T = \omega r$. If the rotating body undergoes an angular acceleration, that is if the rate of rotation speeds up or slows down, then v_T also increases or decreases. In other words, v_T undergoes a linear acceleration called the tangential acceleration. Tangential acceleration is represented as a_T.

There are two equations for tangential acceleration. The first is just a revision of the equation for linear acceleration from Chapter 5, modified to specify the focus on v_T, and is written as follows:

$$a_T = \frac{v_{Tf} - v_{Ti}}{v_t}$$

The second equation calculates a_T on the basis of α, the angular acceleration of the body and r, the radius of rotation. This equation is $a_T = \alpha r$ and requires α to be measured in rad/s/s.

Centripetal acceleration

Centripetal acceleration is related to the change in direction of v_T as a body rotates. Centripetal acceleration is represented as a_c. As long as the body rotates, a_c is present, even if the body rotates at a constant ω, without any angular acceleration. An initial response to this definition might be, "Huh?", but let me explain.

Linear velocity is a vector quantity. Linear velocity changes, or accelerates, if the body speeds up or slows down *in a particular direction.* The direction of v_T is constantly changing as a body rotates. Look at Figure 9-7. The vector representing v_T is shown at *a*, the moment of release, and also at *b*, some time before release. Note the different orientation of the vector v_T at *a* and *b*. The direction of v_T has changed, so v_T has undergone acceleration.

The direction of this acceleration is always toward the axis of rotation, the center of the circular arc of motion. This leads to the name centripetal acceleration because *centripetal* means "center seeking." The acceleration direction is toward the shoulder for the pitcher.

Centripetal acceleration is an instantaneous acceleration — it is not calculated over a time period. There are two equations to calculate a_c: one based on how fast the body is rotating (ω), and one based on the magnitude of v_T. The equations are

$$a_c = \omega^2 r$$
$$= \frac{v_T{}^2}{r}$$

Chapter 10

Causing Angular Motion: Angular Kinetics

* *

In This Chapter

▶ Determining the moment of inertia

▶ Working with angular momentum

▶ Applying Newton's three laws to angular motion

▶ Relating angular impulse and angular momentum

* *

Torque is the turning effect of a force, and it's created when an external force is applied around an axis of rotation. Most movement consists of individual body segments rotating at joints, and whole body rotations like swings, spins, and twists around an axis. Knowing that torque causes rotation and understanding how torque affects rotation are important for improving performance and reducing the risk of injury.

This chapter begins with an explanation of the *moment of inertia,* the term used to specify the resistance of a body to changing angular motion. I go on to explain angular momentum as the product of the body's moment of inertia and its angular velocity, and I show how manipulating the moment of inertia of a body affects its angular velocity even in the absence of an external torque. Newton's laws of motion, which I explain in reference to linear motion in Chapter 6, are shown to give a better understanding of how torque causes the rotation of a body. Finally, I show how a torque applied for a period of time affects the angular momentum of a body using the angular version of the impulse–momentum relationship.

Resisting Angular Motion: The Moment of Inertia

The resistance of a body to a change in *linear motion* is called *inertia,* and it depends on the mass of the body. The linear inertia of a body is the same for a change in motion in any direction because mass stays constant.

The resistance to a change in *angular motion,* or *rotation,* is called the *moment of inertia.* It's represented with the capital letter *I.* The moment of inertia of a body depends on the mass of the body and how the mass is distributed around an identified axis of rotation. It can be different depending on the axis of rotation around which the body can rotate, as explained below. The greater the moment of inertia, the greater the resistance to changing angular motion.

Individual segments rotate at the joints, such as bending and straightening the elbow joint, and the whole body can rotate, such as performing a somersault, cartwheel, or pirouette. In this section, I explain the moment of inertia of an individual body segment and the moment of inertia of the whole body.

The moment of inertia of a segment

Consider the bending and straightening of your elbow. Each particle of mass in your forearm contributes to the moment of inertia of the forearm around the joint (elbow) axis. The moment of inertia of the segment is the sum of the moments of inertia of all the mass particles. Stated mathematically, that's

$$I_{\text{segment}} = I_1 + I_2 + I_3 + \cdots + I_n$$

where I_{segment} is the moment of inertia of the segment, its resistance to changing angular motion, and I_1 to I_n are the moments of inertia created by the individual particles, with n being the total number of particles in the segment.

The contribution of an individual particle is affected by the mass of the particle and its distance — how far it is — from the axis of rotation. Consider three individual particles A, B, and C, randomly chosen from all the particles that make up the thigh, as indicated in Figure 10-1. The identified axis of rotation is a mediolateral axis at the hip joint (the axis around which flexion and extension of the thigh occur).

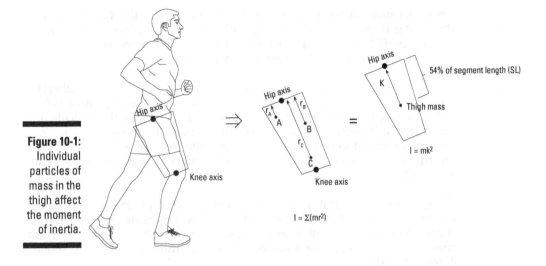

Figure 10-1:
Individual particles of mass in the thigh affect the moment of inertia.

The moment of inertia of an individual particle depends on the mass (m) of the particle and its distance (r) squared from the identified axis of rotation, as described in the equation $I = mr^2$. Substituting this equation for the contribution of an individual particle into the equation to calculate the moment of inertia of the thigh as a whole gives the following:

$$I_{thigh} = I_A + I_B + I_C \cdots + I_n$$
$$I_{thigh} = (m_A r_A^2 + m_B r_B^2 + m_C r_C^2 + \cdots + m_n r_n^2)$$

which can be reduced to

$$I_{thigh} = \Sigma(m_i r_i^2)$$

where I represents the individual particles.

The units for the moment of inertia are the kg · m², reflecting the unit of mass, the kilogram (kg), multiplied by the unit of distance, the meter, squared (m²).

The equation $I = \Sigma(m_i r_i^2)$ shows that a more massive particle and a particle farther from the axis both create more resistance to angular motion. The distance has a greater effect because of r^2 in the equation.

In the thigh in Figure 10-1, if particles A, B, and C are all the same mass, the I from particle C is greater than the I from particle B, which is greater than the I from particle A, because particle C is the farthest from the axis and particle A is the closest.

Obviously, using $\Sigma(m_i r_i^2)$ to calculate the moment of inertia of a segment like the thigh gets very difficult. Any segment in a living body has millions (at least) of particles, and the particles are made of different materials, such as bone, muscle, and fat, which have different masses.

An alternative value, the *radius of gyration,* is used to simplify calculating the moment of inertia of a segment. The symbol for the radius of gyration is k. Using the radius of gyration eliminates the need to consider the individual particles in a segment. The radius of gyration basically considers all the mass of a segment to be located at a single point from the axis. Think of k as an "equivalent distance" for the mass distribution in a segment, as shown in Figure 10-1.

Using the radius of gyration, the moment of inertia of a segment depends on the total mass (m) of the segment and its radius of gyration (k) squared from the identified axis of rotation, as described in the equation $I = mk^2$. Note that there is no summing in this equation, because all the mass acts at one point, k, from the axis.

To calculate the radius of gyration for a segment, you need two values:

✔ The length of the segment (in meters).

✔ The distance from the specified axis of rotation to the segment's radius of gyration. This value is calculated using the segment length and a table specifying the location of k as a proportion of the segment length.

Consider the thigh in Figure 10-1. Note that the thigh, like many segments in the human body, is cone shaped — the diameter is greater at the end near the hip (the proximal end) than at the end near the knee (the distal end). The mass of the thigh is distributed closer to the thigh axis compared to the knee axis.

Most segments in the extremities are cone shaped because most of the muscle bellies are located near the proximal end and most of the long tendons from these muscles are located at the distal end.

Table 10-1 shows the location of k as a proportion of the segment length from both the proximal and the distal end of the segment. The values are all for calculating k around the mediolateral axis. For the thigh, k is located 54 percent ($0.540 \times 100\%$) of the segment length from the hip axis and 65.3 percent ($0.653 \times 100\%$) of the segment length from the knee axis, as shown in Figure 10-1.

Table 10-1	The Radius of Gyration/Segment Length around a Mediolateral Axis	
Segment	*Proximal End*	*Distal End*
Upper arm	0.542	0.645
Forearm	0.526	0.647
Thigh	0.540	0.653
Lower leg	0.528	0.643

The proportion (p) from Table 10-1 is used with the actual segment length (SL) to calculate the radius of gyration (k) using the equation $k = \text{SL} \times p$. If the length of the thigh is 30 centimeters (cm), or 0.3 meters, here are the steps to calculate the location of k around the mediolateral axis at the proximal end (the hip joint):

1. **Identify the known values and create a table of variables.**

   - k_{hip} = Radius of gyration, to be solved

   - SL = Segment length = 0.3 m

   - p_{hip} = Location of k as a proportion of segment length from the proximal end = 0.54

2. **Select the equation and substitute the given values.**

   $k = \text{SL} \times p = 0.3 \text{ m} \times 0.54 = 0.162 \text{ m}$

The radius of gyration of the thigh is located 0.162 m from the hip joint.

To calculate k from the distal end of the thigh (the knee joint), follow the same steps:

1. **Identify the known values and create a table of variables.**

   - k_{knee} = Unknown, to be solved

   - SL = 0.3 m

   - p_{knee} = 0.653

2. **Select the equation and substitute the given values.**

   $k = \text{SL} \times p = 0.3 \text{ m} \times 0.653 = 0.196 \text{ m}$

The radius of gyration of the thigh is located 0.196 m from the knee joint.

To calculate the moment of inertia (I) of the segment, use the calculated value for k and the mass (m) of the segment in the equation $I = mk^2$. I'll set the mass (m) of the thigh to be 8 kg and show the steps to calculate the I of the thigh around the hip (proximal end of the segment) and then around the knee (distal end of the segment).

To calculate I of the thigh around a mediolateral axis through the hip joint:

1. **Identify the known values and create a table of variables.**
   - I_{hip} (moment of inertia) = Unknown, to be solved
   - m (mass of the thigh) = 8 kg
   - k_{hip} = Radius of gyration around the hip joint = SL $\times p$ = 0.162 m

2. **Select the equation and substitute the given values.**

   $I = mk^2 = 8 \text{ kg} \times 0.162 \text{ m}^2 = 8 \times 0.026 = 0.21 \text{ kg} \cdot \text{m}^2$

The moment of inertia of the thigh around the hip joint is 0.21 kg · m².

To calculate I of the thigh around a mediolateral axis through the knee joint, follow the same steps:

1. **Identify the known values and create a table of variables.**
   - I_{knee} = Unknown, to be solved
   - m = 8 kg
   - k_{knee} = Radius of gyration around the knee joint = SL $\times p$ = 0.196 m

2. **Select the equation and substitute the given values.**

   $I = mk^2 = 8 \text{ kg} \times 0.196 \text{ m}^2 = 8 \times 0.038 = 0.31 \text{ kg} \times \text{m}^2$

The moment of inertia of the thigh around the knee joint is 0.31 kg · m².

The moment of inertia of the thigh is smaller when measured to the hip joint axis than when measured to the knee joint axis. This makes sense, because more of the mass of the thigh is located closer to the hip joint than it is to the knee joint (as shown in the p values for each and because the thigh is cone shaped). The distance of the mass from the axis of rotation is the greatest influence on the moment of inertia.

I showed the calculation of the moment of inertia in two steps: calculate k, and then calculate I. The two steps can be combined using the equation $I = m(\text{SL} \times p)^2$.

The moment of inertia of the whole body

In many movements, like a somersault, pirouette, or cartwheel, the whole body rotates as a single unit around an imaginary axis. The moment of inertia of the whole body around the axis depends on how far the individual segments like the arm, leg, and head are positioned from the axis of rotation. Although the mass of a body segment doesn't change, segments can be moved closer to or farther from the axis of rotation.

The equation for the moment of inertia of the whole body around an axis can be considered as $I_{axis} = \Sigma(m_{seg}r_{seg}^2)$, where m_{seg} is the mass of an individual segment and r_{seg} is the distance from the segment to the axis of rotation. Because the distance is squared, the positioning of the segments has a greater effect on I_{axis} than does the mass of the segment.

Figure 10-2 shows four body positions, two relative to a mediolateral axis (A and B) and two relative to a vertical axis (C and D). The effect of the position of individual segments on the moment of inertia is shown in the approximate *I* values listed below each figure. Two main points from Figure 10-2 are

✔ **The moment of inertia is greater when segments are positioned farther from the axis of rotation compared to when segments are closer to the axis of rotation.** The moment of inertia around the ML axis is almost four times greater when the arms are held overhead in an extended position (A, *I* about 15 kg · m²) compared to when the body is in the tucked position (B, *I* about 3.5 kg · m²). Similarly, the moment of inertia around the vertical axis is about two times greater when the arms are held away from the body (C, *I* about 2.2 kg · m²) compared to when the arms are held at the sides (D, *I* about 1.1 kg · m²).

✔ **To reduce the moment of inertia, bring segments closer to the axis of rotation. To increase the moment of inertia, move segments away from the axis of rotation.**

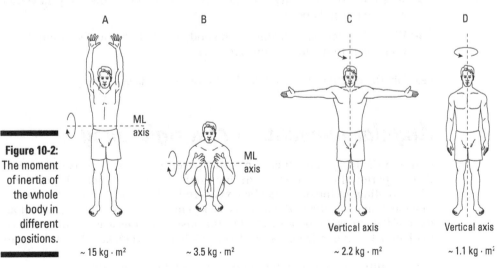

Figure 10-2: The moment of inertia of the whole body in different positions.

The moment of inertia is greater around the mediolateral axis than around the vertical axis. Because the vertical axis runs the length of the body while the mediolateral axis runs sideways across the body, the mass of the individual segments is closer to the vertical axis than to the mediolateral axis, regardless of the positioning of the segments around each axis.

The moment of inertia of the whole body is calculated using the parallel-axis theorem, not simply $I = \Sigma(m_{seg}r_{seg}^2)$. The equation for the parallel-axis theorem is $I_{axis} = \Sigma(I_{seg} + [m_{seg}r_{seg}^2])$, where I_{axis} is the moment of inertia of the whole body around a specific axis, I_{seg} is the moment of inertia of an individual segment about the same specific axis within the segment itself, and r_{seg} is the distance from the axis within the segment to the axis around which the whole body rotates. Pretty complicated, and the simpler equation of $I_{axis} = \Sigma(m_{seg}r_{seg}^2)$ is an acceptable approximation.

Considering Angular Momentum

Angular momentum (H) is the product of the moment of inertia (I) and the angular velocity (ω; see Chapter 9) of a rotating body. As an equation, $H = I\omega$. The units are kg · m²/s (read "kilogram meters squared per second"). The units of angular velocity must be expressed in radians per second, or rad/s (1 radian = 57.3 degrees; see Chapter 9 for more on the radian) to calculate angular momentum.

The calculation of angular momentum differs in two situations:

✔ When the rotating body maintains a fixed position so the body moves as a whole (a rigid body)

✔ When the segments making up the body rotate individually within the body, as is frequent in many activities

I explain the calculation of H in both situations in this section.

Angular momentum of a rigid body

Consider the body in the tucked position (refer to Figure 10-2), like when a diver is performing somersaults during a dive. The diver holds this position while rotating, so the diver can be considered a rigid body during this time. The moment of inertia of the diver is 3.5 kg · m², and I'll set the angular velocity at 600 degrees/s (sounds high, but this instantaneous ω is not uncommon in diving). Use the following steps to calculate the diver's angular momentum:

1. **Identify the known values and create a table of variables.**

   • H_d (angular momentum of the diver) = Unknown, to be solved

- I_d (moment of inertia of the diver) = 3.5 kg · m²
- ω_d (angular velocity of the diver) = 600 degrees/s = 10.5 rad/s (converted as 600 ÷ 57.3)

2. **Select the equation and substitute the given values.**

$$H_d = I_d \omega_d = 3.5 \text{ kg} \cdot \text{m}^2 \times 10.5 \text{ rad/s} = 36.8 \text{ kg} \cdot \text{m}^2\text{/s}$$

The angular momentum of the diver is 36.8 kg · m²/s.

Angular momentum of the human body when individual segments rotate

A diver typically leaves the board in the extended position, shown in Figure 10-2a, and then rotates the individual segments to get into the tucked position shown in Figure 10-2b. Each rotating segment contributes to the H of the diver in two ways:

✔ **Local contribution:** The local contribution is created by the individual segment as it rotates around its own axis, or H_{local}. The equation is $H_{local} = I_{seg} \times \omega_{seg}$. (This is the calculation of H for the segment as if it's an individual rigid body.)

✔ **Remote contribution:** The remote contribution comes from the segment as it rotates relative to the axis of the whole body, or H_{remote}. The equation is $H_{remote} = \text{mass}_{seg} \times r_{SAWA}^2 \times \omega_{SegWBaxis}$, where mass_{seg} is the mass of the individual segment, r_{SAWA} is the distance from the segment axis to the whole body axis, and $\omega_{SegWBaxis}$ is the angular velocity of the segment around the whole body axis.

The H of the diver is calculated as the sum of the local and remote components from each of the segments using the equation $H = \Sigma(H_{local} + H_{remote})$.

The total H of a body is higher when the local and remote contributions of each segment are higher. The total H of the body is the sum of the H around each of the axes around which the body is rotating. For example, if a diver is somersaulting around the mediolateral axis and twisting around the vertical axis, the H of the diver is the sum of the $H_{ML} + H_V$.

A New Angle on Newton: Angular Versions of Newton's Laws

Newton's laws of motion (see Chapter 6) explain that an unbalanced force causes an acceleration of a body, meaning the magnitude of the acceleration is proportional to the magnitude and direction of the unbalanced force and inversely proportional to the body's mass, and that forces equal in magnitude and opposite in direction are applied to the two bodies interacting to produce the force. In most human motion, the mass of the body stays constant, and only the motion of the body can change.

Torque (see Chapter 8) is the turning effect of a force. If an applied force does not pass through a specified axis of rotation, the force has a moment arm around the axis and creates a torque on the body. When more than one torque acts around an axis, calculating the net torque at the axis (ΣT_{axis}) shows if an unbalanced torque ($\Sigma T_{axis} \neq 0$; see Chapter 8 for more on calculating net torque) is present around the axis. Newton's laws of motion explain the effect of an unbalanced torque on the angular motion of a body.

The two main points to remember in looking at Newton's laws for angular motion are that the moment of inertia — the resistance to changing angular motion — is not necessarily constant for a body, and a force must have a moment arm to produce a turning effect on a body. In the following sections, I explain Newton's laws of angular motion.

Maintaining angular momentum: Newton's first law

Newton's first law, also known as the law of conservation of angular momentum, specifies that an unbalanced external torque causes a change in the angular momentum of a body. Angular momentum is the product of the moment of inertia (I) and the angular velocity (ω), or $H = I\omega$ (as explained earlier). According to Newton's first law:

> A body will continue to rotate about its axis of rotation with a constant value of angular momentum unless acted on by an unbalanced external torque.

The H of a rotating body stays constant in the absence of an unbalanced external torque. But the angular motion of a rotating body can change without an external torque.

Angular motion is a description of how fast and in what direction a body is rotating, its angular velocity (ω). The resistance to changing angular motion is the moment of inertia (I). Even though $H = I\omega$ stays constant without an external torque, because I can be changed by repositioning the mass around an axis of rotation, the angular velocity of a body can change even without an unbalanced external torque. This is different from the linear situation, where a body accelerates, or speeds up or slows its linear motion, only because of an unbalanced external force; the mass stays constant.

The angular velocity (ω) of a body does *not* have to stay constant in the absence of an unbalanced torque. It's the angular momentum (H), the product of the moment of inertia (I) and the angular velocity (ω), that is conserved in the absence of an unbalanced torque.

Some of the clearest examples of Newton's first law for angular motion come in events that include one or more total-body rotations while the performer is in the air. Diving is an example. Figure 10-3 shows a diver's body positions at several points during the performance of a one-and-one-half front somersault, from when the diver leaves the board (a) until the diver's hands contact the water (f). Below the body positions, I've plotted out the diver's momentum (H), moment of inertia (I), and angular velocity (ω) while in the air.

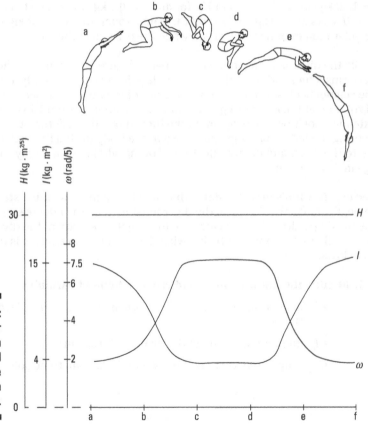

Figure 10-3: Angular momentum is conserved while the diver is in the air.

Between leaving the board and contacting the water, the diver is a projectile (see Chapter 5). The only external force acting on the diver is body weight, applied through the diver's center of gravity (at about waist level; see Chapter 9). All the forces within the diver are internal forces, and they do not affect the motion of the diver as a projectile (more on internal forces in Chapter 4).

While the diver is in the air, the center of gravity is the axis of rotation for the front somersault. Because the diver's body weight is applied at the center of gravity, it's a centric force on the diver and creates no external torque to change the angular momentum of the diver. This situation is the same for any performer in the air and considered a projectile — there are no external torques applied. Without any external torques on a projectile, the angular momentum stays at the same value throughout the time in the air, in accordance with Newton's first law for angular motion.

Figure 10-3a shows the diver leaves the board in an extended position, with $I = 15.0$ kg $\cdot$ m^2 and $\omega = 2.0$ rad/s, for an $H = 30.0$ kg $\cdot$ m^2/s. The figure shows that H stays at 30.0 kg $\cdot$ m^2/s until the hands touch the water at Figure 10-3f. Angular momentum is conserved while the diver is in the air.

While the diver is rising in the air (between Figure 10-3a and 3c), the diver performs a tuck, bringing the arms and legs in close to the body. By bringing the segments closer to the body, the mass of the diver is closer to the axis of rotation at the center of gravity and the I decreases from 15 kg $\cdot$ m^2 in the extended position to 4.0 kg $\cdot$ m^2 in the tucked position. All the muscle torques to bring the individual segments close to the body during the tuck are internal to the diver, and do not affect the H of the whole diver. H stays at 30.0 kg $\cdot$ m^2/s while tucking.

Because Newton's first law states that angular momentum must stay the same, the angular velocity of the diver must increase as I decreases during the tucking action. You can calculate the angular velocity of the diver when in the fully tucked position by knowing that angular momentum is conserved, as follows:

1. **Identify the known values and create a table of variables.**

   - H (angular momentum) = Constant at 30.0 kg $\cdot$ m^2/s (Newton's first law)

   - I_d (moment of inertia of the diver) = 4.0 kg $\cdot$ m^2

   - ω_d (angular velocity of the diver) = Unknown, to be solved

2. **Select the equation $H = I\omega$, isolate for the unknown, and substitute the given values.**

$$\omega = \frac{H}{I} = \frac{30 \text{ kg} \cdot \text{m}^2/\text{s}}{4.0 \text{ kg} \cdot \text{m}^2} = 7.5 \text{ rad/s}$$

The angular velocity of the diver increases to 7.5 rad/s while the diver is in the tucked position. This angular velocity stays at 7.5 rad/s as long as the diver stays in the tucked position, keeping *I* at 4.0 kg · m², as shown in Figure 10-3 between c and d. Angular momentum stays at 30.0 kg · m²/s.

To enter the water in a head-first, extended position with the arms over-head, the diver untucks. All the muscle torques to reposition the body to the extended position are internal to the diver. Just as when the diver was tuck-ing, these internal torques don't affect the *H* of the whole diver.

During the untucking, the *I* increases as the masses of the segment are moved away from the axis of rotation. Because *I* increases, ω decreases. Consider the position of the diver between positions d and e while untucking. As *I* increases from 4.0 kg · m² at d to 9.0 kg · m² at e, ω decreases from 7.5 rad/s to 3.2 rad/s.

To calculate the *I* of the diver when *H* is constant and ω is known, say between e and f when ω is 2.2 rad/s, use the following steps:

1. **Identify the known values and create a table of variables.**

    - *H* (angular momentum) Constant at 30.0 kg · m²/s (Newton's first law)

    - I_d (moment of inertia) = Unknown, to be solved

    - ω_d (angular velocity) = 2.2 rad/s

2. **Select the equation $H = I\omega$, isolate for the unknown, and substitute the given values.**

$$I = \frac{H}{\omega} = \frac{30 \text{ kg} \cdot \text{m}^2/\text{s}}{2.2 \text{ rad/s}} = 13.6 \text{ kg} \cdot \text{m}^2$$

The moment of inertia must be 13.6 kg · m² for the diver to have an angular velocity of 2.2 rad/s.

Angular momentum is conserved when no external torque acts on a projec-tile. Once in the air, no more angular momentum can be gained, and no angu-lar momentum can be lost. The diver can use only internal muscle torques to reposition the segments of the body and manipulate *I,* which causes a change in the angular velocity of the diver's rotating body.

Changing angular momentum: Newton's second law

Newton's second law for angular motion explains what an unbalanced external torque will do to the angular momentum of a body:

> An *unbalanced* external torque applied to a body causes a change in the angular momentum of the body with a magnitude proportional to the unbalanced torque, in the direction of the unbalanced torque, and inversely proportional to the body's moment of inertia.

Mathematically, the law can simply be stated as $\Sigma T_{axis} = I_{axis}\alpha_{axis}$, where ΣT_{axis} is the net torque (the sum of the external torques on the body) around a specified axis, I_{axis} is the moment of inertia of the body around the specified axis, and α_{axis} is angular acceleration of the body around the specified axis. An unbalanced torque means that the net torque is not 0, or $\Sigma T_{axis} \neq 0$ (see Chapter 8). An unbalanced external torque causes an angular acceleration of the body. This changes the angular velocity (ω), which changes the angular momentum ($I\omega$).

The equation $\Sigma T_{axis} = I_{axis}\alpha_{axis}$ relates an unbalanced torque to angular acceleration, but the law states an unbalanced torque causes a change in the angular momentum of a body. To clear this up, substitute the mathematical definition of angular acceleration, $\alpha = \frac{\Delta\omega}{\Delta t}$ (see Chapter 9), into the equation:

$$\Sigma T = I\alpha$$
$$= I\frac{\Delta\omega}{\Delta t} = \frac{\omega_f - \omega_i}{\Delta t} = \frac{I\omega_f - I\omega_i}{\Delta t}$$

This shows that an unbalanced torque changes the angular momentum ($I\omega$) of a body. I'll come back to unbalanced torque causing a change in angular momentum later in the chapter.

Figure 10-4 shows a diver during the push-off from a platform to perform a forward front somersault. The moment of inertia (I) around the center of gravity (C of G) of the diver (the axis of rotation) is 15 kg · m² in this position. The diver is pushing down and back on the platform, and the platform pushes upward ($F_V = 550$ N) and forward ($F_H = 50$ N) on the diver. The moment arm (MA) of each force around the center of gravity is shown on the figure (MA is explained in Chapter 8). The MA for F_V is 0.5 m and the MA for F_H is 0.6 m. To calculate the angular acceleration (α) of the diver around the center of gravity, I'll calculate the net torque around the diver's center of gravity and use this value in the equation $\Sigma T_{axis} = I_{axis}\alpha_{axis}$ to calculate the angular acceleration. Start by calculating the net torque (ΣT_{CofG}) on the diver, using $T = F \times \text{MA}$ to calculate the torques from the F_V and F_H (see Chapter 8):

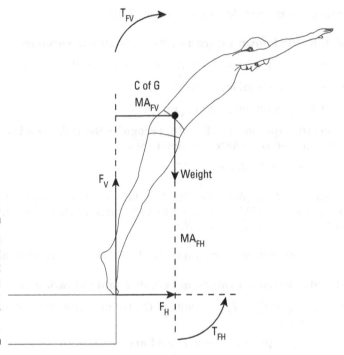

Figure 10-4: The external forces at the diver's feet cause an angular acceleration.

Although the diver's body weight is an external force, it's applied at the center of gravity of the diver, which is the axis of rotation. Body weight is centric to the axis of rotation, so it doesn't create a torque around the center of gravity.

1. **Identify the known values and create a table of variables.**

   - T_{FV} (torque from the vertical force) = Unknown, to be solved
   - F_V (vertical force) = 550 N
   - MA_{FV} (moment arm of F_V) = 0.5 m

2. **Select the equation $T = F \times \text{MA}$, recognize the unknown is already isolated, and substitute the given values.**

   $$T = (+550 \text{ N})(0.5 \text{ m}) = 275 \text{ Nm}$$

The torque from F_V is 275 Nm, tending to rotate the diver forward in the clockwise (CW) direction. The CW direction is the negative direction for torque, so the $T_{FV} = -275$ Nm.

Use the same steps to calculate the torque from F_H:

1. **Identify the known values and create a table of variables.**
   - T_{FH} (torque from the horizontal force) = Unknown, to be solved
   - F_H (horizontal force) = 20 N
   - MA_{FV} (moment arm of F_V) = 0.6 m

2. **Select the equation T = F × MA, recognize the unknown is already isolated, and substitute the given values.**

   $T = (+20 \text{ N})(0.6 \text{ m}) = 12 \text{ Nm}$

The torque from F_H is 12 Nm, tending to rotate the diver backward in the counterclockwise (CCW) direction. The CCW direction is the positive direction for torque, so the $T_{FH} = +12$ Nm.

Calculate the angular acceleration of the diver around the center of gravity:

1. **Identify the known values and create a table of variables.**
   - $\Sigma T_{CofG} = T_{FV} + T_{FH,}$ the sum of the torques around the center of gravity
   - T_{FV} (torque from the vertical force) = −275 Nm
   - T_{FH} (torque from the horizontal force) = +12 Nm
   - I_{CofG} (moment of inertia around the center of gravity) = 15 kg · m^2
   - α_{CofG} (angular acceleration around the center of gravity) = Unknown, to be solved

2. **Select the equation $\Sigma T_{CofG} = I_{CofG}\alpha_{CofG}$, isolate the unknown, substitute the given values, and solve.**

$$\alpha_{CofG} = \frac{\Sigma T_{CofG}}{I_{CofG}} = \frac{T_{FV} + T_{FH}}{I_{CofG}}$$
$$= \frac{(-275 \text{ Nm}) + (+12 \text{ Nm})}{15 \text{ kg} \cdot \text{m}^2} = \frac{-263 \text{ Nm}}{15 \text{ kg} \cdot \text{m}^2} = -17.5 \text{ rad/s/s}$$

The angular acceleration of the diver around the center of gravity in this position during the pushoff is –17.5 rad/s/s. The negative sign indicates the CCW, or forward, direction of the angular acceleration. Note that angular acceleration is measured in rad/s/s, and not degrees/s/s.

Nm/kg · m^2 leads to rad/s/s as follows:

$$\frac{\text{Nm}}{\text{kg}\cdot\text{m}^2} = \frac{(\text{kg}\cdot\text{m/s/s})\cdot\text{m}}{\text{kg}\cdot\text{m}\cdot\text{m}} = \frac{\text{kg}\cdot\text{m}\cdot\text{m/s/s}}{\text{kg}\cdot\text{m}\cdot\text{m}}$$

The kg · m · m cancel out, leaving /s/s. Inserting radian (rad) as a dimensionless quantity relating linear measures to angular measures (see Chapter 9) leads to rad/s/s.

Equal but opposite: Newton's third law

An eccentric force occurs when the line of action of an external force does not pass through a specified axis of rotation and the external force creates a torque around the axis (as explained in Chapter 8). When the external force is a muscle force creating torque at a joint, the torque is equal in magnitude and opposite in direction on the segments meeting at the joint. Newton's third law for angular motion summarizes the description of the torques on the two bodies, or segments:

> When a single external force creates a torque on two bodies that share an axis of rotation, the torque on one body is equal in magnitude and opposite in direction to the torque on the other body.

Consider the cheerleader shown in Figure 10-5. The cheerleader jumps straight up into the air, leaving the ground in an upright position (a) and bends forward to touch the toes (b). The upper and lower parts of the cheerleader's body move toward each other during the movement as the cheerleader "bends" forward at the indicated axis. Both the upper and the lower body rotate.

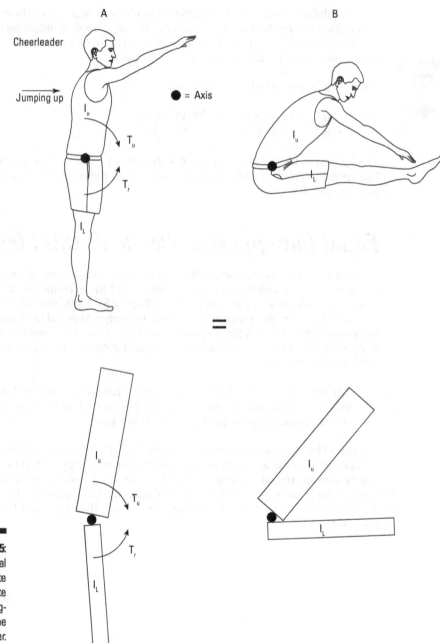

The link segment model in the lower part of Figure 10-5 shows how Newton's third law applies to the cheerleader. I've drawn a rectangle to represent the upper body and a rectangle to represent the lower body (a very simple link segment model). The different sized rectangles represent the moments of inertia (I). I of the upper body is larger than I of the lower body, because there is more mass and it's farther away from the axis of rotation.

The muscles crossing the front of the hips create a torque on both segments of the cheerleader. The torque applied to the upper body is equal in magnitude but opposite in direction to the torque applied to the lower body. The torque on the upper body causes it to rotate in the clockwise (CW) direction. The torque on the lower body causes it to rotate in the counterclockwise (CCW) direction.

The effect of the equal but opposite torques is seen in the change of position of each segment. The upper body, with a larger I, does not rotate as far in the clockwise (CW) direction as the lower body, with a smaller I, rotates in the counterclockwise (CCW) direction.

Newton's third law should not be learned as "when one body exerts a torque on another, the torques are equal and opposite." This simplification overlooks that torque is created by an external force acting eccentrically to an axis. The torques are only equal when the two bodies share an axis of rotation and the external force creates a torque at the shared axis.

Changing Angular Momentum with Angular Impulse

Earlier, I show that the equation $\Sigma T = I\alpha$ can be rewritten as

$$\Sigma T = \frac{I\omega_f - I\omega_i}{\Delta t}$$

If both sides of the equation are multiplied by Δt, the equation becomes $\Sigma T \Delta t = I\omega_f - I\omega_i$. This equation is the impulse–momentum relationship for angular motion. The left side of the equation, $\Sigma T \Delta t$, is called *angular impulse*. Angular impulse is the product of an external unbalanced torque multiplied by the duration of time it's applied. The units of angular impulse are Newton · meter · seconds, or N · m · s (the units of force multiplied by the units of moment arm [m] multiplied by units of time). Angular momentum is the product of a body's moment of inertia and its angular velocity, as explained earlier in this chapter. The right side of the equation, $I\omega_f - I\omega_i$ (read as "final angular momentum minus initial angular momentum"), describes a change in the angular momentum of a body. The equation shows that an applied angular impulse can increase or decrease the angular momentum of the body over the time period.

The equation is similar to the impulse–momentum relationship for linear motion explained in Chapter 6. The big difference is that both I and ω can change while the angular impulse is applied, while in linear motion the mass stays constant and only the linear velocity can change.

Chapter 11

Fluid Mechanics

A fluid is a substance that flows. Both water (a liquid) and air (a gas) are fluids. Fluids are important in mechanics because the molecules of a fluid interact with a body immersed or moving in the fluid. Both air and water create buoyant, lift and drag forces, and these are pushing forces on a body. Water flowing over a swimmer in the pool pushes on the swimmer. Air flowing over a pitched baseball pushes on the ball. Water flowing over the arms of an exerciser in a water aerobics class pushes on the arms. Air flowing over a ski jumper pushes on the jumper.

Fluid forces typically resist the motion of a body, but manipulation of the lift, drag, and buoyant forces can benefit performance. Fluid forces affect both linear and angular motion, but I cover fluid mechanics in this part of the book because some of the most interesting applications of fluid force are to manipulate rotations. These rotations include how a human floating in water tends to sink feet first and how moving balls can curve.

This chapter begins with a description of buoyant force, the pushing force present when a body is immersed in a fluid. Next, I explain the drag and lift forces created when a body moves through a fluid. Finally, I describe two special cases of fluid force — known as the Bernoulli principle and the Magnus effect.

Buoyancy: Floating Along

Buoyant force is the force pushing up on a body immersed in a fluid. As a body gets immersed in a fluid, the body pushes down on the fluid and the fluid pushes up on the body. Although the fluid exerts an upward force at every point of contact with the body, the resultant buoyant force can be considered to be applied at a single point, the *center of buoyancy*.

Figure 11-1a shows a person floating on the surface of a swimming pool. To be in the water, the body must have pushed the water out of the way. (Think about it — when you get into a tub of water, you displace or push the water out of the way.) When a body pushes on the water, the water pushes back on the body (as explained by Newton's third law; see Chapter 3). The force of the water, the buoyant force, represents an external force applied to the body. If the buoyant force from the fluid is large enough, it causes the body to float. The entire body of the person is the focus of the analysis; Figure 11-1b is a free-body diagram showing the external forces acting on the body. Drawing a free-body diagram is explained in Chapter 4.

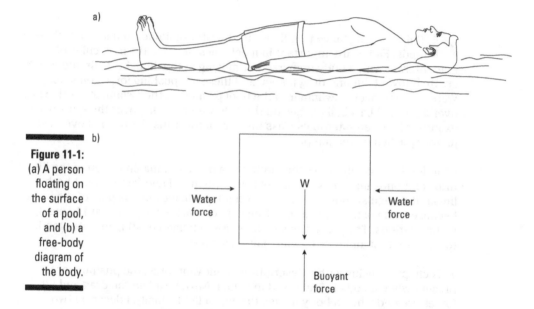

Figure 11-1:
(a) A person floating on the surface of a pool, and (b) a free-body diagram of the body.

A body floating in a fluid is described as at rest in the *vertical* (up and down) direction. A body at rest in the vertical direction isn't accelerating; there's no acceleration because the external forces acting in the vertical direction are balanced (as explained by Newton's first law; see Chapter 3).

The weight of the swimmer's body acts, as weight always acts, downward in the vertical direction. But there are no unbalanced vertical forces, so some vertical force must be offsetting the weight pulling down on the body. The force that pushes up on a body in the water is termed the *buoyant force,* and it's the only possible source of this vertical force. Because the water force pushing to the left on the swimmer is offset by the water force pushing to the right on the swimmer, there is no unbalanced external force in the horizontal direction.

Not all objects, and not all people, will float. Whether a body floats or sinks is determined by the magnitude of the body's weight relative to the buoyant force that can be created. If the weight of the body is greater than the buoyant force that can be created, the body sinks. If the weight of the body is less than or equal to the buoyant force that can be created, the body floats.

So, what determines the magnitude of the buoyant force relative to the weight of the body? The buoyant force comes from the water pushing back on the body as the body pushes it out of the way. The amount of water displaced is determined by the volume of the body — the three-dimensional space occupied by the body (the body's height × width × length for a rectangular body, or $\frac{4}{3}\pi r^3$ for a round body). A unit to express volume is cubic meters (m^3). The actual magnitude, or size, of the buoyant force is the weight of the volume of the fluid displaced. Think about it — water has weight. A jug full of water weighs more than the same jug with no water in it. Displacing a larger volume of water creates a larger buoyant force.

Technically, an empty jug isn't actually empty. It's full of air, another liquid, but air weighs so much less than water that we call the jug "empty," even when it isn't. Aren't we foolish sometimes?

Archimedes' principle states that the magnitude of the buoyant force on a body is equal to the weight of the volume of the fluid the body displaces.

But will the buoyant force be large enough to counteract the weight of the body? The weight of the body, of course, is the pull of gravity on the mass of the body (see Chapter 4). A human body contains various materials, including bone, muscle, fat, air (in the lungs and gut), and water (the basis of the cell). Each has a different weight — bone is heavier than muscle, which is heavier than water, which is heavier than fat, which is heavier than air.

Specific gravity relates the weight of a material to its volume, and provides a quick reference to see if a material will float in a fluid. Water, which weighs 9,800 Newtons per cubic meter (N/m^3), is assigned the value of 1.0 as its specific gravity. Comparing the weight of a cubic meter of other materials to the weight of a cubic meter of water gives each material a specific gravity value. Table 11-1 presents the weight per cubic meter and specific gravity values for air, fat, water, muscle, and bone.

Table 11-1	The Weight and Specific Gravity for Components of the Human Body	
Material	**Weight**	**Specific Gravity**
Air	11.8 N/m^3	0.0012
Fat	8,829 N/m^3	0.9
Water	9,800 N/m^3	1.0
Muscle	10,300 N/m^3	1.05
Bone	9,800–14,715 N/m^3	1.0–1.5

To determine whether a material will float in a fluid, compare the specific gravities. If the material's specific gravity value is less than the fluid, the material will float in the fluid. The values for air (0.0012) and fat (0.9) show that they both float in water. The values for muscle (1.05) and bone (1.0 to 1.5) show that they both sink in water.

Typically, the specific gravity of the intact human body is around 1.06, because most of our body weight comes from bone and muscle. Because 1.06 is greater than 1.0, people tend to sink in water. But 1.06 is measured with minimal air in the lungs. A deep breath adds more air, decreasing specific gravity to less than 1.0 and making floating easier. Keeping air in the lungs to reduce the specific gravity is a good reason to relax while in a swimming pool.

Individuals differ in their ease of floating, because the proportion of bone, muscle, and fat (called *body composition*) varies. Sex, race, age, and physical activity level all affect body composition. With training, almost everyone can learn to float, mainly by learning to hold air in and relax. A "sinker" can always go to the Dead Sea, which has a high specific gravity of 1.16 because of the high salt content — bad for fish, good for floating.

Relative density is another way to relate the volume of a body to a similar volume of water. Density is mass/volume, but because weight is proportional to mass (see Chapter 4), relative density and specific gravity provide similar comparative values of a material to water.

To make my point, I lied to you in Figure 11-1. For most people, the buoyant force and the weight of the body are not aligned when a body floats horizontally in the water. Figure 11-2a shows a better representation of a human body floating horizontally. Because the lungs are filled with air, the center of buoyancy is located at about the level of the lungs, offset from the center of gravity of the human body, which is around the level of the *navel* (belly button).

The offset of the weight and the buoyant force creates a torque on the body, rotating the feet downward and causing the body to sink. To prevent this, the center of gravity and the center of buoyancy must be aligned. If the floater bends at his knees and holds his arms out to the side, as in Figure 11-2b, the center of gravity will move toward the head and become colinear with the center of buoyancy. Now, no torque is created, and the body will float.

a)

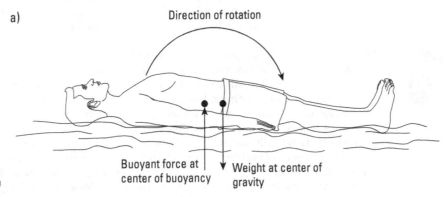

Direction of rotation

Buoyant force at center of buoyancy   Weight at center of gravity

Figure 11-2: (a) A better representation of a body floating horizontally, and (b) the effect of bending the knees and holding the arms out.

b)

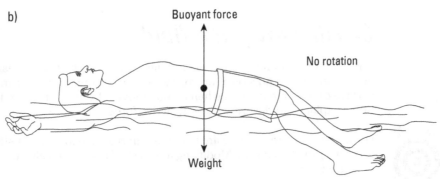

Buoyant force

No rotation

Weight

Considering Force Due to Motion in Fluid

When a body moves through a fluid such as water or air, the body pushes on the fluid, and the fluid pushes on the body, in accordance with Newton's third law (see Chapter 3). Because the body and fluid are moving, the forces are called *dynamic fluid forces*. Two effects occur as a moving body interacts with the fluid:

✔ The fluid must be redirected to flow around the moving body; this effect creates *form drag* and *lift*.

✔ The fluid passing over but in direct contact with the surfaces of the body interacts with the surfaces to create a force called *surface drag*.

In this section, I describe how form drag, surface drag, and lift affect the motion of the body, and I explain the factors affecting each force. Finally, I explain Bernoulli's principle and the Magnus effect, special cases of lift force created by the shape of the body and the spin of the body.

Dynamic fluid forces are present when a body is moving through a stationary fluid, such as when swimming laps in a pool, when the body is stationery and the fluid is moving (such as swimming in a *flume* [a sort of "treadmill" for swimmers]), or when both the body and the fluid are moving (such as when swimming back to shore against an outgoing tide). *Relative velocity* describes how fast the fluid and the body are moving relative to each other, regardless of which body is actually moving. If a swimmer moves through stationary water at 0.8 meters per second (m/s), the relative velocity is 0.8 m/s. If the water in a swimming flume is moving at 0.8 m/s over a swimmer stationary in the flume, the relative velocity is 0.8 m/s. If a swimmer moves forward at 0.5 m/s into a head-on tide moving toward her at 0.3 m/s, the relative velocity is 0.8 m/s.

Causing drag in a fluid

The word *drag* implies something holding you back. *Drag forces* resist a body moving in a fluid and slow it down. Two types of drag force — form drag and surface drag — are created when a body moves through a fluid. In this section, I explain and describe these resistive forces.

Form drag and surface drag are two different resistive forces, but similar factors affect the two forces. Modifying any of the factors typically affects both resistive forces.

Form drag

As a body moves through a fluid, the fluid gets pushed out of the way by the body. Of course, as the body pushes on the fluid, the fluid pushes back on the body, too. Form drag is the component of the fluid force pushing on the body opposite to the direction of the body's motion through the fluid, resisting the motion of the body. (I explain the other component of the fluid force, called *lift*, a little later in this chapter.)

The front of the body moving through the fluid causes the fluid to separate by pushing on it. The fluid pushes backward on the body, resisting its forward motion. After the fluid separates, it flows around the body before rejoining at the back of the body. How easily the fluid flows over the body, and how easily the fluid rejoins at the back of the body, affects a pocket of swirling fluid, called *turbulence,* created around and behind the body. The turbulent fluid applies a forward push to the body, assisting its forward motion. Greater turbulence produces a smaller forward push. The difference between the backward push as the fluid separates and the forward push as the fluid flows and rejoins is the *form drag.* Obviously, the larger the backward push as the fluid separates, and the smaller the forward push as the fluid flows and rejoins, the larger the resultant form drag. In this section, I explain the specific factors affecting form drag, beginning with the two bodies moving through a fluid depicted in Figure 11-3.

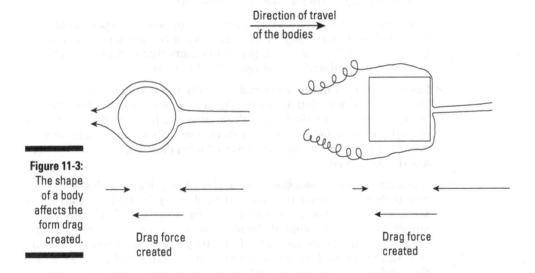

Direction of travel
of the bodies

Drag force
created

Drag force
created

Figure 11-3:
The shape
of a body
affects the
form drag
created.

In the left side of Figure 11-3, the round body has a small effect on the fluid. A relatively small force is required to deflect the fluid away from the curved front surface of the body. The resisting force backward on the body is small. The fluid flows smoothly over the round edges of the body and rejoins relatively easily at the back of the body. Only a small pocket of turbulence is created. There is little difference between the backward push and the forward push on the body, creating minimal form drag.

In the right side of Figure 11-3, the square body has a large effect on the fluid. A relatively large force is required to deflect the fluid away from the flat front surface of the body. The resistive force on the body is large. The fluid does not flow smoothly over the sharp top and bottom edges of the body, disrupting the flow, and the disrupted fluid does not rejoin easily at the back of the body. A large pocket of turbulence is created, producing a forward push that is much smaller than the backward push. A lot of form drag is created.

In essence, form drag is affected by how easily the fluid *deflects,* or separates, from the front of the body and by how easily the fluid rejoins after passing over the body. The factors influencing the deflection and rejoining of the fluid around a moving body include the following:

- ✔ **Fluid viscosity:** *Viscosity* refers to the ease with which a fluid flows — sort of a measure of the "thickness" of the fluid. Fluids with greater viscosity flow less easily than fluids with lower viscosity. Greater form drag is created when a body moves in a more viscous fluid, because moving a thicker fluid requires more force. Water is more viscous than air, which means more form drag is created in water than in air.

  Fluid viscosity is affected by temperature. The water in a pool used for high-level swimming competitions such as the Olympics is maintained between 25°C (77°F) and 28°C (82°F) to ensure that performances are not aided, or inhibited, by the viscosity of the water.

- ✔ **Relative velocity of the body and the fluid:** Greater form drag is created with a higher relative velocity. In fact, the square of the relative velocity (v^2) contributes to form drag. However, because going fast is an objective of most performances, reducing velocity is not a useful way to reduce form drag. It's just accepted that form drag will be greater when a body moves faster.

- ✔ **The shape and cross-sectional area of the body:** Figure 11-3 shows how body shape affects the flow of the fluid over the body. The cross-sectional area relates to the width of the front surface of the body that initially runs into the fluid. Reducing cross-sectional area reduces form drag. This is the factor related to form drag over which there is the most control. Streamlined body positions (such as a crouched position over the handlebars of a bicycle) and aerodynamic equipment design (such as tight-fitting running clothes) are examples of modifying shape and cross-sectional area to reduce form drag during performance.

Surface drag

Surface drag is the force exerted on a body by the molecules of fluid flowing across — you guessed it — the surface of the body in motion through the fluid. When a fluid flows across the surface of the body, the molecules of the fluid contact any bumps, or irregularities, in the surface texture. The "collisions" between the fluid molecules and the bumps exert resistive force on the body, and the sum of all the collision forces is the surface drag. Greater surface drag is produced by larger and/or more collision forces.

The factors affecting the size and number of the collisions include the following:

- ✔ **Fluid viscosity:** Forces are larger when a more *viscous,* or thicker, fluid collides with the surface bumps. The mass of the individual molecules in a fluid affects its viscosity. Water is more viscous than air because water molecules are more massive than air molecules. Compared to air molecules, the more massive molecules of water exert a greater force on the surface bumps during a collision. Larger collision forces, and greater surface drag, are created when a body moves in a more viscous fluid.

- ✔ **Surface texture:** More collisions are created when the body surface is rough. The bumps of a rough surface jut out farther into the fluid, and more molecules of the fluid collide with the larger bumps. More collision forces, and greater surface drag, are created when a fluid flows over a rough surface.

Of these two factors, a performer has the most control over surface texture. To smooth the body surface, some swimmers and runners shave their body hair before competing. Modern materials, manufactured with an extremely smooth surface to reduce interaction with the fluid molecules, are used to make clothing that fits tightly to the body of the performer. This clothing includes bodysuits for swimmers or skiers and even jerseys for hockey players.

Causing lift in a fluid

Unlike form drag and surface drag, *lift force,* or *lift,* is not a resistive force. The word *lift* implies something raising a body up, but the lift force created as a body moves through a fluid can push up or down on the body.

Lift, like form drag, is created when a body moving through a fluid pushes the fluid out of the way. Form drag, a resistive force, is the component of the fluid force that acts opposite to the direction of motion. Lift is the component of the fluid force acting perpendicular to the direction of the drag force.

The orientation of the body in the fluid affects the direction of the lift force. The term *angle of attack* describes the orientation of the body to the fluid in which it moves. No angle of attack indicates that the body travels flat. A positive angle of attack indicates that the body is tilted with the front of the body higher than the back. A negative angle of attack indicates that the body is tilted but with the front of the body lower than the back. Figure 11-4 shows three bodies moving through a fluid at different angles of attack, represented with θ, and the direction of the lift and form drag forces created.

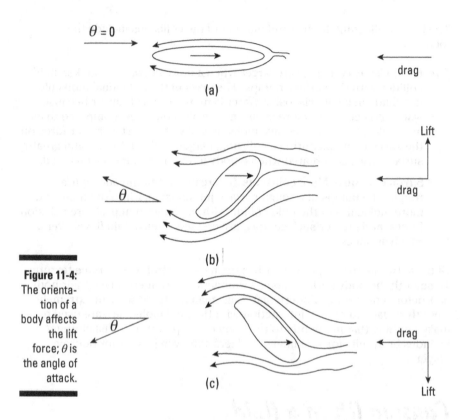

$\theta = 0$

drag

(a)

Lift

drag

(b)

Figure 11-4:
The orienta-
tion of a
body affects
the lift
force; θ is
the angle of
attack.

θ

drag

(c)

Lift

In Figure 11-4a, the body moves directly forward into the fluid, as indicated by the velocity vector drawn in the body. The body is level, so it has no angle of attack ($\theta = 0$) to the fluid. The fluid separates equally over and under the body; all the fluid force on the body is drag.

In Figure 11-4b, the body moves forward and upward into the fluid, as indicated by the velocity vector drawn in the body. The body has a positive angle of attack. The fluid separates unequally over the body because the bottom surface has more exposure to the oncoming fluid. The bottom surface of the body pushes forward and downward on the fluid, while the fluid pushes backward and forward on the body (explained by Newton's third law; see Chapter 3). The fluid force pushing backward is the form drag; the fluid force pushing upward is the lift force. With a positive angle of attack, lift pushes up on the body.

In Figure 11-4c, the body also moves forward and upward into the fluid, as indicated by the velocity vector drawn in the body, but the body has a negative angle of attack. As in Figure 11-4b, the fluid separates unequally over the body, but with a negative angle of attack, the top surface has more exposure to the oncoming fluid. The top surface of the body pushes forward and upward on the fluid, while the fluid pushes backward and downward on the body. The fluid force pushing backward is the form drag; the fluid force pushing downward is the lift force. With a positive angle of attack, lift pushes down on the body.

The resultant fluid force acts directly opposite the velocity of the body. The resultant fluid force can be resolved (see Chapter 3) into the drag and lift components using trigonometry. This shows that drag acts backward to resist motion, and lift acts either up or down. Drawing the resultant of the drag and lift forces in Figure 11-4 will verify this idea.

Bernoulli's principle and the Magnus effect are examples of lift created by differences in fluid velocity on opposite sides of a body. The Bernoulli principle is related to the shape of the body, while the Magnus effect is related to the spin of the body.

Bernoulli's principle

Individual molecules in a fast-moving fluid exert less force on the molecules beside them than the individual molecules in a slow-moving fluid exert on the molecules beside them. Because of this, lower pressure is present in a fast-moving fluid than is present in a slow-moving fluid. This effect, known as the *Bernoulli principle,* is the basis for using the aerodynamic shape of an airfoil to create lift. An airfoil has two long surfaces aligned in the direction of the fluid flow, as shown in Figure 11-5. One of the long surfaces (in Figure 11-5, the top surface) is more curved than the opposite surface (in Figure 11-5, the bottom surface).

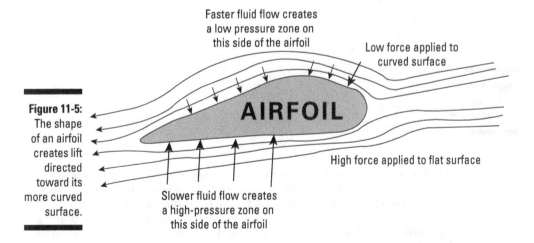

Figure 11-5: The shape of an airfoil creates lift directed toward its more curved surface.

The fluid flowing across the top, curved surface travels faster than the fluid flowing across the flatter, bottom surface. The difference in the speeds of the fluid flow creates low pressure above the curved surface and high pressure below the flat surface. The low pressure of the fluid above the curved surface pushes down on the airfoil with less force than is applied up on the airfoil by the high pressure of the fluid below the flat surface. The overall result is a lift force applied up on the airfoil.

If the airfoil is flipped over, so the curved surface is on the bottom and the flat surface is on the top, the lift force is directed down on the airfoil.

The Magnus effect

The *Magnus effect* states that lift force can be created by the spin of a round body. Lift is created because spin causes the relative velocity between the round body and the fluid to be different on opposite sides of the body.

Figure 11-6 shows a ball traveling to the right and spinning clockwise. Relative to the center of the ball, the air molecules are flowing over the ball from right to left. But note that the spin of the ball causes point A to be moving to the right and point B to be moving to the left, relative to the center of the ball. An air molecule striking point A is slowed down a lot because point A is moving in the opposite direction, while an air molecule striking point B is not slowed down as much because point B is moving in the same direction. In accordance with the Bernoulli principle, the slower air moving over the side of point A exerts more force on the ball than the faster air moving over the side of point B. The Magnus effect creates a lift force acting down on the ball from point A to point B.

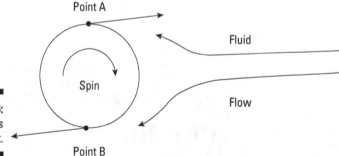

Figure 11-6: The Magnus effect.

The design of some balls accentuates the Magnus effect. The laces on a baseball create a more vertical surface, increasing how much the air slows down on the surface of the ball striking the oncoming air. Dimples on a golf ball can have the same effect. Skilled pitchers and golfers intentionally apply spin to the ball to cause a larger lift force and alter the path of the ball through the air.

Part IV
Analyzing the "Bio" of Biomechanics

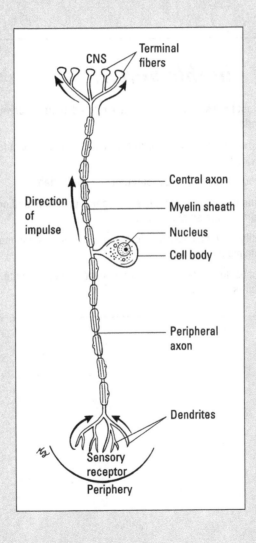

Find out about the types of motor units in a free article at http://www.dummies.com/extras/biomechanics.

In this part...

- ✔ Find out how stress and strain are related in the mechanics of material.

- ✔ Discover the anatomical framework of the neuromusculoskeletal system.

- ✔ Use bones for support, protection, and movement.

- ✔ Outline the nervous system as the communication and control system for human movement.

- ✔ Identify how the microscopic protein filaments within muscle create muscle tension.

- ✔ Use muscle and tendon for effective movement with the stretch-shorten cycle.

Chapter 12

Stressing and Straining: The Mechanics of Materials

*H*ave you ever sprained your ankle or at least "tweaked" it? What do you think the response of the ligaments was during the sprain? Did they change in length? Did some of the ligaments' fibers get so stretched that they ruptured? Which is easier to stretch: a ligament or a bone?

Mechanics of materials looks at questions like these. Specifically, material mechanics deals with how the "stuff" (material) making up a body responds to the external forces applied to it. Applying an external force is referred to as *loading* the body. The applied load creates internal forces on the material within the body, called *stress*. The stress causes the materials within the body to deform or change length, a response called *strain*.

For example, when your foot turns inward at the ankle, the ligaments undergo stress. The stress causes the ligaments to change length, or experience strain. How much strain develops determines whether a ligament gets sprained.

Because stress and strain are within the body, as internal forces, the stress is ignored when looking at how an external force tends to change the linear and rotary motion of the body. (You can find more detail on the effect of external forces on linear and angular motion in Chapters 6 and 10, respectively.)

The stress and strain within the body caused by loading are the focus of this chapter. I begin this chapter by explaining the concept of stress. I define the types of stress imposed by different external forces applied to the body. Then I explain the concept of strain and the different deformations caused by different types of stress. I outline the stress–strain relationship and define important descriptors of the relationship. This chapter gives you an overview of the basic concepts of the mechanics of materials needed to understand how loading affects a body.

An entire book, *Mechanics of Materials For Dummies,* by James H. Allen III, PE, PhD (Wiley), is devoted to an in-depth coverage of the concepts introduced in this chapter.

Visualizing Internal Loading of a Body

An internal force is created within the body when an external force is applied. (See Chapter 4 for an explanation of internal and external forces.) To calculate the internal force within a body, an analysis plane is identified at a point within the body. The *analysis plane* is an imaginary slice through the body at a point of interest.

Figure 12-1 shows a solid cylinder made of a single material. Two colinear forces F_1 (+1,000 N) and F_2 (–1,000 N) are pushing in opposite directions on the ends of the cylinder. The cylinder is at rest (it's in equilibrium; see more on equilibrium in Chapter 8), so the sum of the forces is 0. Mathematically, $\Sigma F = F_1 + F_2 = 0$.

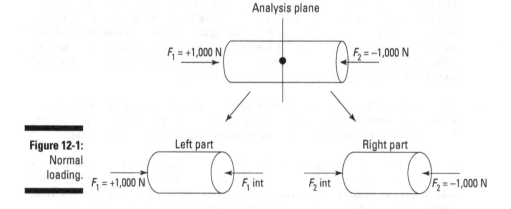

Figure 12-1:
Normal
loading.

I've drawn an analysis plane at the center of the cylinder in Figure 12-1. When the cylinder is separated into two parts at the analysis plane, each of the parts is in equilibrium. The internal forces F_{1int} and F_{2int} act on each of the separate parts at the analysis plane. The internal force on one part of the cylinder comes from its contact with the part of the cylinder on the other side of the analysis plane.

Focusing on the left part of the cylinder, F_{1int} is the pushing force applied at 90 degrees to the analysis plane surface of the left part by the analysis plane surface of the right part. F_{1int} and F_{2int} are equal and opposite forces applied to each part of the separated cylinder. The forces are internal to the body, so they aren't included in a free-body diagram of the cylinder when the cylinder is the focus of the analysis.

To calculate the internal force F_{1int}, use $\Sigma F = 0$ as follows:

$$\Sigma F = 0 = F_1 + F_{1int}$$

Now isolate for F_{1int}:

$$F_{1int} = -F_1$$

Now substitute in the known value and solve:

$$F_{1int} = -(+1{,}000 \text{ N}) = -1{,}000 \text{ N}$$

The internal force on the left part of the cylinder at the analysis plane is $-1{,}000$ N. F_{1int} is applied at 90 degrees to the surface of the analysis plane and is called a *normal load.*

Figure 12-2 shows the same cylinder at rest with different external forces applied. The forces F_1 ($-1{,}000$ N) and F_2 ($+1{,}000$ N) are pushing in opposite directions on the sides of the cylinder and the forces are not colinear. Mathematically, $\Sigma F = F_1 + F_2 = 0$.

I've separated the cylinders at the analysis plane. Focusing on the left part of the cylinder, F_{1int} is the pushing force applied on the analysis plane surface by the right part. The half-tip to the vector for F_{1int} and F_{2int} indicates these forces are applied along the surface of the analysis plane, not at 90 degrees.

To calculate the internal force F_{1int}, use $\Sigma F = 0$ as follows:

$$\Sigma F = 0 = F_1 + F_{1int}$$

Isolate for F_{1int}:

$$F_{1int} = -F_1$$

Substitute in the known value and solve:

$$F_{1int} = -(-1{,}000 \text{ N}) = +1{,}000 \text{ N}$$

The internal force on the left part of the cylinder at the analysis plane is $+1{,}000$ N. F_{1int} is applied along the surface of the analysis plane and is called a *shear load.*

TIP

Shear forces are identified with vectors that have half-tips indicating the direction of the force.

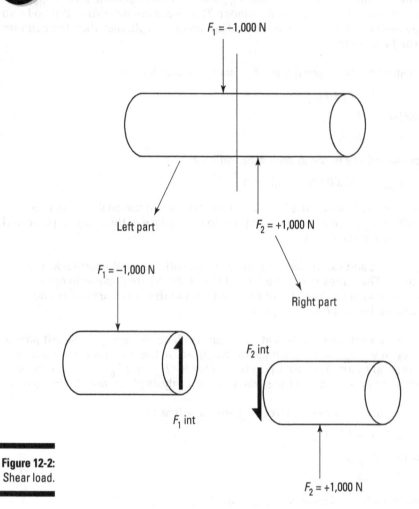

$F_1 = -1,000$ N

Left part

$F_2 = +1,000$ N

$F_1 = -1,000$ N

Right part

F_2 int

F_1 int

Figure 12-2:
Shear load.

$F_2 = +1,000$ N

Applying Internal Force: Stress

Stress is a measure of how the internal force is distributed across the internal structure of the body. The internal force (F_{int}) is assumed to be evenly distributed over the cross-sectional area (A) of the analysis plane on which it's applied. As an equation,

$$\text{Stress} = \frac{F_{int}}{A}$$

The unit of stress is the Pascal (Pa), and 1 Pa = 1 N/m² of stress.

Stress measures internal force per unit area, $\frac{F_{int}}{A}$.

For a given force, stress is higher if the area is small than if the area is large. Stress is inversely proportional to the area over which the force is applied. No one wants to get stepped on, but if you had to choose between being stepped on by someone wearing high heels or that same person wearing sneakers, which would you choose? You'd choose the sneaker scenario because the stress would be smaller because sneakers have a larger area.

For a given area, stress is higher if the force is large than if the force is small. Stress is directly proportional to the magnitude, or size, of the force. No one wants to get stepped on, but if you choose between getting stepped on by a child or an adult, which would you choose? You'd choose the child, because the stress would be lower because the child probably weighs less (a smaller force).

Stress is typically reported in a unit called the MegaPascal (MPa), equal to 1,000,000 Pa. Few structures in the human body have cross-sectional areas measured in meters, so typically area is measured in cm² (centimeters squared). With F_{int} measured in N and A measured in cm², a little conversion of units is necessary to report stress in MPa. Using 100 cm = 1 m, and 1 MPa = 1,000,000 Pa, here are the steps for conversion:

$$\left(\frac{F_{int} \text{ in N}}{A \text{ in cm}^2} \right) \times \left(\frac{100 \text{ cm}}{m} \right)^2 \times \left(\frac{1 \text{ MPa}}{1,000,000 \text{ Pa}} \right) = \text{Stress in MPa}$$

The second part, $\times \left(\frac{100 \text{ cm}}{m} \right)^2$, converts the area measured in cm² to area measured in m², expressing the stress in the appropriate units of N/m² (Pa).

The third part, $\times \left(\frac{1 \text{ MPa}}{1,000,000 \text{ Pa}} \right)$, converts the units of stress from Pa to MPa.

I'll demonstrate the conversion of units when calculating stress below.

There are two major types of stress classified by how the F_{int} is applied to the surface area of the analysis:

- **Normal stress** is created when the F_{int} is a normal load, with the force acting normal (at 90 degrees) to the cross section of the body (normal load is explained earlier).

- **Shear stress** is created when the F_{int} is a shear load, acting parallel to, or along, the cross section of the body (shear load is explained earlier).

In the following sections, I explain the two types of stress and show the different forms of each type. I begin with the three forms of normal stress, then move on to shear stress, including the special case of shear stress created by torsion, or twisting, within the body.

Normal stress

Normal stress occurs when a normal load is applied at the analysis plane. As explained earlier, a normal load means F_{int} is applied at 90 degrees, or normal, to the cross-sectional area. The F_{int} represents the load created within the body by the external force acting on the body. When the external force is applied centrically through an axis of the body, the term *axial loading* is used to describe the imposed load (applying force centrically through an axis is explained in Chapter 8). The F_{int} produced only pushes or pulls on the cross section. The symbol for a normal stress is σ (pronounced sigma). There are two principal types of normal stress (tension and compression), and one combined form of normal stress called *bending,* a combination of tension and compression.

Tension

Tension is the form of loading produced within a body when a pair of external collinear forces pull on a body, as shown in Figure 12-3. The line of action of the external pulling forces passes through an imaginary axis at the analysis plane, as indicated in Figure 12-3. An example of tension is the loading of a tendon when a muscle is active. The muscle fibers pull on the tendon at one end, and the bone the tendon inserts on pulls at the other end.

Tension is an axial load. Both F_{int} act at 90 degrees to the cross-sectional area of the analysis plane. The pulling stress created within the body is termed tensile stress, and is represented with σ.

Because F_{int} at the analysis plane is assumed to be distributed evenly over the cross-sectional area of the analysis plane, the stress at the analysis plane is represented with the multiple force vectors drawn on the analysis plane. The sum of the multiple small vectors is equal to the F_{int} — the individual vectors represent components of F_{int}. The number of vectors I've drawn is only representative of the distribution of F_{int} over the surface area. In reality, there are as many vectors as there are bonds between the molecules of the material at the analysis plane.

The tensile stress created in a body is calculated as $\sigma = \dfrac{F_{int}}{A}$.

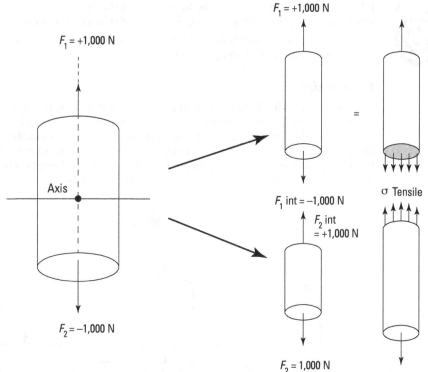

Figure 12-3:
Tensile
stress.

Consider the top portion of the cylinder in 12-3. Solving $\Sigma F = F_1 + F_{1\text{int}} = 0$, $F_{1\text{int}}$ is determined to be $-1,000$ N. If the area of the analysis plane is 4 cm^2, I use the following steps for the calculation of σ:

1. **Create a table of variables.**

   - $F_{1\text{int}} = -1,000$ N

   - $A = 4$ cm^2 (***Note:*** Remember that the basic unit of area is the m^2.)

   - $\sigma =$ Unknown tensile stress, to be calculated in MPa

2. **Select the equation for tensile stress, fill in the known values, and use the conversion steps to express σ in MPa.**

$$\sigma = \frac{F_{1\text{int}}}{A} = \frac{-1,000 \text{ N}}{4 \text{ cm}^2} \times \left(\frac{100 \text{ cm}}{\text{m}}\right)^2 \times \left(\frac{1 \text{ MPa}}{1,000,000 \text{ Pa}}\right) = -2.5 \text{ MPa}$$

The negative sign of the calculated value σ shows that the stress is pulling downward on the analysis plane. Tensile stress represents a pulling load on the internal structure of the body.

TIP

If I'd calculated the stress on the lower part of the cylinder, the calculated stress would be +2.5 MPa because $F_{2int} = +1,000$ N. The stress on one side of the analysis plane is equal in magnitude but opposite in direction to the stress on the other side of the analysis plane.

Compression

Compression is the form of loading produced within a body when a pair of external collinear forces pushes on a body, as shown in Figure 12-4. The line of action of the external pushing forces passes through an imaginary axis at the analysis plane, as indicated in Figure 12-4. An example of compression is the loading of the cartilage covering the end of bones where two bones meet at a joint. The cartilage is squeezed between the two bones.

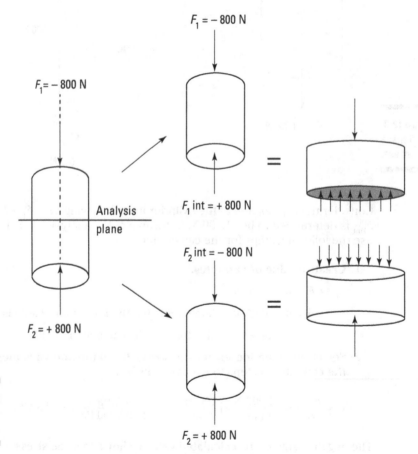

Figure 12-4: Compressive stress.

Compression is an axial load. Both F_{int} act at 90 degrees to the cross-sectional area of the analysis plane. The squeezing stress created within the body is termed compressive stress, represented as σ.

Compressive stress is calculated as $\sigma = \dfrac{F_{int}}{A}$.

Consider the top portion of the cylinder in 12-4. Solving $\Sigma F = F_1 + F_{1int} = 0$, F_{1int} is determined to be +800 N. If the area of the analysis plane is 12 cm², I use the following steps for the calculation of σ:

1. **Create a table of variables.**

   - $F_{1int} = +800$ N

   - $A = 12$ cm²

   - $\sigma =$ Unknown, to be calculated in MPa

2. **Select the equation for compressive stress, fill in the known values, and use the conversion steps to express σ in MPa.**

$$\sigma = \frac{F_{1int}}{A} = \left(\frac{+800 \text{ N}}{12 \text{ cm}^2} \right) \times \left(\frac{100 \text{ cm}}{\text{m}} \right)^2 \times \left(\frac{1 \text{ MPa}}{1,000,000 \text{ Pa}} \right) = +0.67 \text{ MPa}$$

The positive sign of the calculated values σ shows the stress is pushing upward on the analysis plane. Compressive stress represents a squeezing load on the internal structure of the body.

Bending

Bending is a complex form of loading produced within a body when two or more external forces or torques are applied to a body. The complexity of bending arises because the F_{int} at the analysis plane creates torque within the body. (See Chapter 8 for more on torque.) I'll explain bending using two forces, as shown in Figure 12-5.

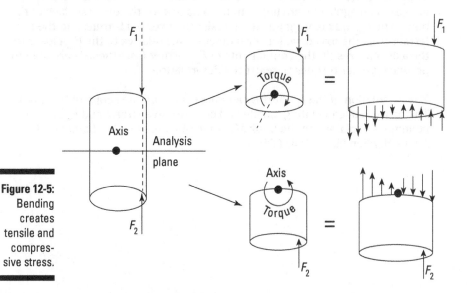

Figure 12-5:
Bending creates tensile and compressive stress.

Bending results from a non-axial load, because the lines of action of the colinear external forces don't pass through the imaginary axis at the analysis plane. Because the external forces are eccentric to the axis in the center of the analysis plane, the forces create torques tending to bend, or curve, the body (hence the name *bending load*) around the imaginary axis. An example of bending is the loading of the femur (thigh bone) when standing upright. The weight of the upper body pushing down on the femur and the force of the tibia pushing up on the femur at the knee joint are eccentric to the long axis of the femur, and they create a bending load.

Separating the cylinder in Figure 12-5 at the analysis plane shows that the torque from F_1 tends to cause the top part to rotate in the clockwise direction, and the torque from F_2 tends to cause the bottom part to rotate in the counterclockwise direction. But because each part of the cylinder is in equilibrium, there are no unbalanced torques ($\Sigma T = 0$) acting on either part of the cylinder. Compression and tension are created within the body on opposite sides of the imaginary axis at the analysis plane to create the torques resisting the turning effect of the external force. There is no normal stress at the axis itself.

Consider the top part of the cylinder in Figure 12-5. The clockwise torque from F_1 is resisted by a counterclockwise torque created at the analysis plane by the F_{int} produced over the surface area. To produce the CCW torque, some components of F_{int} push up on the analysis plane on the same side of the axis as the externally applied load, and some components of F_{int} pull down on the analysis plane on the opposite side of the axis. One side of the surface area shows compressive stress and the other side shows tensile stress.

The magnitudes of the compressive stress and the tensile stress increase from the center of the body toward the outside surface because the F_{int} increase in magnitude moving from the axis toward the outer surface of the body. The F_{int} are creating torque to resist the external torque, and the F_{int} farther from the axis have longer moment arms relative to the F_{int} closer to the axis. As a result, the components of F_{int} farther from the axis are able to produce greater torques to resist the deformation.

The calculation of the F_{int} during bending is beyond the scope of this book. The process of calculating tensile and compressive stress, and F_{int}, during bending is explained in the book *Mechanics of Materials For Dummies*, by James H. Allen III, PE, PhD (Wiley).

Shear stress

Shear stress occurs when a shear load is applied at the analysis plane. As explained earlier, a shear load means F_{int} acts parallel to the cross-sectional area. The F_{int} acts parallel to the surface in response to a pair of external non-colinear forces pushing or pulling on the body in opposite directions. A more complex form of shear stress is created when a twisting load, called *torsion*, is applied to a body. I'll get to the shear stress from torsion after explaining pure shear stress.

Shear stress is produced because the external forces push or pull on the body in opposite directions, as shown in Figure 12-6. Because the external forces don't share the same line of action, the opposing forces tend to slide the two parts of the body across each other at the analysis plane. The F_{int} acts parallel to the analysis plane to resist the sliding, creating shear stress. The symbol for shear stress is τ (tau).

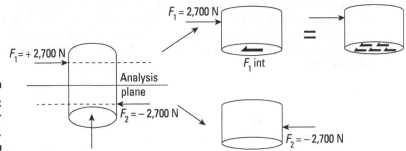

Figure 12-6:
Shear
stress.

Shear stress is calculated as $\tau = \dfrac{F_{int}}{A}$.

Consider the top portion of the cylinder in 12-6. Solving $\Sigma F = F_1 + F_{1int} = 0$, $F1_{int}$ is determined to be –2,700 N. If the area of the analysis plane is 7 cm², I use the following steps to calculate τ:

1. **Create a table of variables.**

   • $F_{1int} = -2{,}700$ N

   • $A = 7$ cm²

   • $\tau =$ Unknown, to be calculated in MPa

2. **Select the equation for shear stress, fill in the known values, and use the conversion steps to express τ in MPa.**

$$\tau = \frac{F_{1int}}{A} = \left(\frac{-2{,}700 \text{ N}}{7 \text{ cm}^2} \right) \times \left(\frac{100 \text{ cm}}{\text{m}} \right)^2 \times \left(\frac{1 \text{ Mpa}}{1{,}000{,}000 \text{ Pa}} \right) = -3.85 \text{ MPa}$$

The negative sign of the calculated τ shows the shear stress is pushing toward the left on the analysis plane. Shear stress represents a load resisting the sliding of the materials within the body.

Tension, compression, and shear are often called the *principal stresses.* Complex forms of loading on a body can usually be resolved into one or more of the principal stresses. More detail on this process is presented in *Mechanics of Materials For Dummies,* by James H. Allen III, PE, PhD (Wiley).

Torsion is produced when a pair of torques is applied in opposite directions around the long axis of a body, as shown in Figure 12-7. The torques tend to cause the body to rotate, or twist, around the axis. An example of torsion is when a person pivots the upper body while keeping the foot planted on the ground. The foot applies a torque on the tibia (bone of the lower leg) in one direction at the ankle, while the femur applies a torque on the tibia in the other direction at the knee joint. At the analysis plane, an internal torque (T_{int}) is present on each surface to resist the external torque. The T_{int} is created by shear forces distributed over the surface area of the analysis plane. Although a shear stress is present on the surface area, as with the calculation of compressive and tensile stresses during bending, the calculation of the shear stress magnitude goes beyond the scope of this book. Again, I refer you to *Mechanics of Materials For Dummies* for a detailed explanation of the calculations.

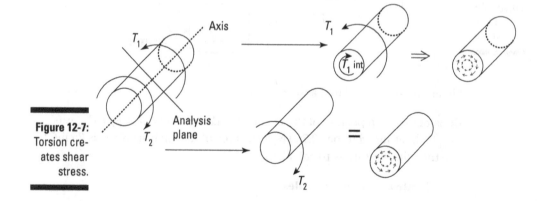

Figure 12-7: Torsion creates shear stress.

When multiple forms of stress are applied at the same time, the body is under combined loading. Because of the complexity of the structure of the human body, and the variety of external forces applied during any form of motion, combined loading is commonly imposed on bones and cartilage. The solution of a combined loading scenario essentially involves identifying and solving for the individual forms of stress. This usually involves making assumptions about the way the various external forces are applied and distributed within the body.

The analysis plane can be set at any point on the body under analysis. In the irregularly shaped structures of the human body such as bones, cartilage, and ligaments, the analysis plane is set at a site that provides useful information to a specific question. The plane can be set at a common site of injury (such as the site of a bone fracture, cartilage injury, or ligament tear).

Responding to Internal Force: Strain

Like all force, stress can't be seen, but the effect of stress can be seen or measured. Stress tends to cause deformation of the body. Deformation means to change shape. Sometimes the deformation is large, like when you push on your skin with your finger. Sometimes the deformation is small, like when you push on your fingernail. Strain is a measure of the deformation of a body in response to the imposed stress. How much a material deforms depends on the type of material and the stress imposed.

Stress *always* causes deformation of a body. Measuring the deformation depends on using an instrument sensitive enough to the amount of deformation expected. Our vision is frequently not precise enough to detect the very small deformations that can occur.

Deformation is measured as a change in length (ℓ) of the body, or $\Delta \ell$. $\Delta \ell$ is the difference between the final length of the body when deformed (ℓ_f) and the initial length of the body (ℓ_i) before deformation. Mathematically $\Delta \ell = \ell_f - \ell_i$. ℓ_i is measured in a specific direction depending on the type of strain to be calculated. The basic equation for calculating strain is as follows:

$$\text{Strain} = \frac{\text{Deformation}}{\text{Initial Length}} = \frac{\ell_f - \ell_i}{\ell_i} = \frac{\Delta \ell}{\ell_i}$$

Strain is a unitless quantity, since $\Delta \ell$ and ℓ_{initial} are measured in the same units (typically meters). The calculated strain value is actually the proportion between the change in length and the initial length. It's common to multiply the calculated proportion by 100 to report strain as a percentage, and that's how I do it.

There are two basic forms of strain:

- **Normal strain:** Normal strain results from the normal stresses of tension and compression. Normal strain is represented with the symbol ε (pronounced epsilon). ε is used for both tensile and compressive strain. Tension and compression change the length and width of the body, as shown in Figure 12-8. Under tension, the stretching deformation of the body involves getting longer and narrower (see Figure 12-8a); the cal-culated $\Delta\ell$ is a positive value, and ε is positive. Under compression, the squeezing deformation of the body involves getting shorter and wider (see Figure 12-8b); the calculated $\Delta\ell$ is a negative value, and ε is negative.

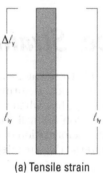

(a) Tensile strain

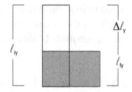

(b) Compressive strain

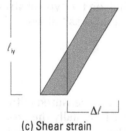

Figure 12-8:
Normal
strain and
shear strain.

(c) Shear strain

✔ **Shear strain:** Shear strain results from a shear stress. Shear strain is represented with the symbol γ (pronounced gamma). Under shear strain, the deformation involves a lateral displacement of the body (see Figure 12-8c), and the amount of lateral displacement represents $\Delta\ell$.

In the following sections, I show the steps to calculate ε for both tensile and compressive loads, and γ for shear loads.

Determining tensile strain

In Figure 12-8a, the initial length is 12 cm. Under the tensile stress, the body elongates to 14 cm. The steps to calculate the tensile strain are as follows:

1. **Create a table of variables.**

   - $\ell_i = 12$ cm
   - $\ell_f = 14$ cm
   - $\varepsilon =$ Unknown tensile strain, to be calculated

2. **Select the equation for tensile strain, fill in the known values, and solve.**

$$\varepsilon = \frac{\text{Deformation}}{\text{Initial Length}} = \frac{\ell_f - \ell_i}{\ell_i} = \frac{14 - 12}{12} = \frac{2}{12} = 0.166 \times 100 = 16.6\%$$

The positive calculated ε indicates the body elongated by 16.6 percent under the tensile load.

Determining compressive strain

In Figure 12-8b, the initial length is 12 cm. Under the compressive stress, the body shortens to 11 cm. The steps to calculate the compressive strain are as follows:

1. **Create a table of variables.**

   - $\ell_i = 12$ cm
   - $\ell_f = 11$ cm
   - $\varepsilon =$ Unknown compressive strain, to be calculated

2. **Select the equation for compressive strain (the same equation used to calculate tensile strain), fill in the known values, and solve.**

$$\varepsilon = \frac{\text{Deformation}}{\text{Initial Length}} = \frac{\ell_f - \ell_i}{\ell_i} = \frac{11 - 12}{12} = \frac{-1}{12} = -0.083 \times 100 = -8.3\%$$

The negative calculated ε indicates the body shortened by 8.3 percent under the compressive load.

Looking at the bodies in Figure 12-8a and Figure 12-8b, the change in shape doesn't occur in only one direction. Under a tensile load, the body gets longer *and* narrower. Think about pulling on a piece of chewing gum (chewed of course!) or a strip of rubber band. As you pull the gum or band, it deforms — it gets longer *and* it gets narrower. Under a compressive load, the body gets shorter *and* wider. Think about stepping on a soda can for this one. When you press down on the can, it deforms — it gets shorter (flattens) and wider (spreads out). The sideways, or width, deformation can also be used to calculate ε_T, the transverse strain (the deformation of the width of the body). Poisson's ratio, represented with the symbol ν (pronounced nu) quantifies the ratio between the strains in the two directions:

$$\text{Poisson's ratio} = \nu = \frac{\varepsilon_{width}}{\varepsilon_{length}}$$

A bigger value of ν reflects a greater deformation in the width of a body relative to the deformation in the length of the body.

Determining shear strain

In Figure 12-8c, the initial length is 12 cm. Under the shear stress, the body laterally displaces 0.1 cm. The steps to calculate the shear strain are as follows:

1. **Create a table of variables.**

   - $\ell_i = 12$ cm

   - $\Delta\ell = 0.1$ cm

   - γ = Unknown shear strain, to be calculated

2. **Select the equation for shear strain, fill in the known values, and solve.**

$$\gamma = \frac{\text{Deformation}}{\text{Initial Length}} = \frac{\Delta\ell}{\ell_i} = \frac{0.1}{12} = 0.008 \times 100 = 0.8\%$$

The calculated γ indicates that the body displaced laterally by 0.8 percent under the shear load.

Straining from Stress: The Stress–Strain Relationship

The *stress–strain relationship,* as the name implies, describes the relationship between the stress imposed on a material and the strain resulting from the imposed load. Typically, a material testing instrument is used to apply known stresses to a prepared specimen of the material while measuring the strain. The values of the stress imposed and the strain measured are plotted, with stress on the vertical axis and strain on the horizontal axis. Joining the points creates a curve showing the relationship between stress and strain for the material. Figure 12-9 shows the curve of a conceptual stress–strain relationship. I refer to Figure 12-9 to identify important aspects of the curve used to describe any material, beginning with the elastic and plastic regions.

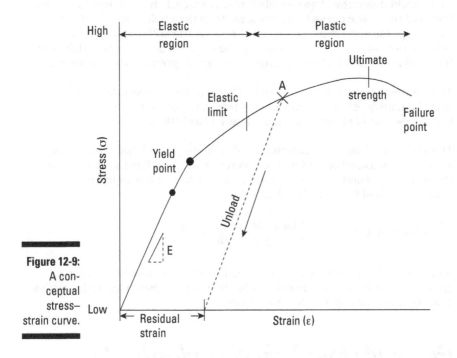

Figure 12-9: A conceptual stress–strain curve.

Give and go: Behaving elastically

Elastic means that a deformed material will return to its original shape when the load causing the deformation is removed. The elastic region is the region of the stress–strain curve where the material behaves elastically. The elastic region extends from 0 stress to the elastic limit (refer to Figure 12-9). The elastic limit indicates the maximal stress and strain a material can withstand before it no longer exhibits elastic behavior. As long as the imposed stress is less than the stress value at the elastic limit, the deformation of the material is temporary and the material returns to its original shape when the stress is reduced. As the load is reduced, the material returns to its original shape following the original stress–strain curve.

The yield point (refer to Figure 12-9) is a point in the elastic region showing where the strain begins to increase more with an increase in stress. Up to the yield point, the stress–strain curve is essentially a straight line. After the yield point, the curve is always less steep than before the yield point. The steepness of a stress–strain curve reflects the stiffness, or resistance to deformation, of a material. Essentially, the material becomes less stiff after the yield point, and it changes length more for a given increase in stress.

The stiffness of a material refers to its ability to resist deformation. A stiff material undergoes little strain when stress is imposed. A pliant material undergoes a lot of strain — a pliant material has little stiffness.

The elastic modulus (E), sometimes called Young's modulus of elasticity, is used to compare the stiffness of different materials. E is the slope of the straight line portion of the stress–strain curve before the yield point (see Figure 12-9), and it is calculated as follows:

$$\text{Elastic Modulus} = E = \frac{\text{Change in Stress}}{\text{Change in Strain}} = \frac{\Delta\sigma}{\Delta\varepsilon}$$

A higher E value indicates a stiffer material, one that deforms less for a given increase in stress. A lower E value indicates a pliant material, one that deforms more for a given increase in stress.

Give and stay: Behaving plastically

Plastic means that a deformed material does not return to its original shape when the load causing the deformation is removed. The plastic region is the region of the stress–strain curve where the material behaves plastically. The plastic region extends from the elastic limit to the failure point.

If the level of stress and strain exceeds the elastic limit, the internal alignment of the material is rearranged, or damaged. Point A on the curve in Figure 12-9 reflects a loading of the material into the plastic region. As the body is unloaded, the stress–strain curve is altered from its shape before being loaded into the plastic region. This is shown with the dashed line from Point A to the level of 0 stress. After loading into the plastic region, a material exhibits greater strain at a level of stress than before it was damaged. The material remains deformed as shown by the residual strain present even when no stress is imposed.

The ultimate strength (refer to Figure 12-9) refers to the maximum stress the material is able to withstand. The ultimate strength of a material is in the plastic region, and the damage within the material from the strain at this level of imposed stress causes greater rearrangement of the internal structure. After the ultimate strength of a material is exceeded, the strain actually begins to decrease.

The failure point of the material is the point at which the material actually physically separates, called *tearing, rupturing,* or *fracturing.* Because of this loss of structural integrity, strain is no longer developed in response to any stress (it can't develop because there aren't any contact points within the material). The material has failed — hence, the name *failure point.*

A few clarifications on the stress–strain relationship as it relates to biological materials. Biological materials can heal. This means the internal arrangement can be somewhat restored, to a different degree in different biological tissues. For example, following complete rupture of a ligament, the internal structure of the ligament cannot be restored through the healing process. However, with proper treatment, the fracture site in a bone can heal, restoring the structural integrity of the bone. Some textbooks refer to a material being "permanently deformed" if it's loaded into the plastic region. The definition of *permanent deformation* doesn't take into account healing within a biological tissue, so I purposely avoid this term.

Chapter 13

Boning Up on Skeletal Biomechanics

. . . .

In This Chapter

▶ Understanding the roles that bones play

▶ Looking at what bones are made of

▶ Focusing on the shapes of bones

▶ Modeling and remodeling bones

. . . .

*T*he skeletal system includes the bones, the joints where two bones come together, and the cartilage and ligaments at the joints. This chapter *doesn't* focus on the names of the approximately 206 bones, 360 joints, and 900 ligaments of the human body. Instead, skeletal biomechanics looks at how the skeletal system provides the general form of the body, provides support, and allows for movement. The skeletal system is very dynamic, affecting and responding to interaction with the variety of forces encountered during a lifetime. The skeletal system is more complex than the space shuttle, and it's even more amazing in its ability to adapt and repair itself.

Skeletal biomechanics describes how the structure and architecture of bones and the ligaments and cartilage at joints affect and are affected by support and movement of the body. This chapter also describes how the skeletal system is affected by mechanical stress, which can be good (for example, when training) or bad (for example, when injuries or diseases occur).

If you're interested in the names of individual bones, joints, and joint actions, *Anatomy & Physiology For Dummies,* 2nd Edition, by Maggie Norris and Donna Rae Siegfried (Wiley), is a good starting point.

What the Skeletal System Does

The skeletal system has five main roles:

- **Blood cell production:** Most of the red and white cells that make up blood are produced inside the bones.

- **Mineral storage:** Bones contain many minerals, mostly calcium. Calcium provides strength to the bone. Calcium is also used by muscles in the process of creating tension, or muscle force, and in the nervous system to transmit signals between nerves. Bones provide a reservoir of calcium for use when adequate calcium is not readily available for use by muscles and nerves. Calcium is withdrawn from the bones to make up for the shortfall. However, withdrawing too much calcium can lead to a decrease in bone strength.

- **Protection of critical organs:** These critical organs include

  - The central nervous system (the brain is protected by the skull, and the spinal cord is protected by the vertebrae of the spinal column, also known as the backbone, which is actually 26 separate bones)

  - Internal organs, like the heart, lungs, liver, and spleen, all of which are protected by the rib cage

  - The kidneys, bladder, and reproductive organs, which are protected by the bones of the pelvis

- **Allowance for movement:** The bones provide a system of rigid *levers* (links) controlled at the joints by the torques produced by muscle forces. The torque created by muscles at the joints causes, prevents, and controls the rotation of the linked segments to allow interaction with the environment. Chapter 8 explains how muscle force creates a *torque,* or turning effect, at a joint.

- **Support of the body when at rest and when moving:** The bones and ligaments are pulled on by muscles to counter the always present force of gravity pulling down on the body. Without the skeletal system, just standing upright would be impossible.

The last three of these roles are mechanical roles of the skeletal system.

How Bones Are Classified

There are two main divisions to the skeletal system (as shown in Figure 13-1):

- **Axial skeleton:** The axial skeleton is the central part, or axis, of the skeleton. The axial skeleton includes the skull, rib cage, and bones of the spine.

✔ **Appendicular skeleton:** The appendicular skeleton is attached to, or appended to, the axial skeleton. The appendicular skeleton includes the arms and the legs, as well as the *scapulae* (shoulder blades; each bone is a scapula) and pelvis. The scapulae are the links between the arms and the axial skeleton, while the pelvis links the legs to the axial skeleton.

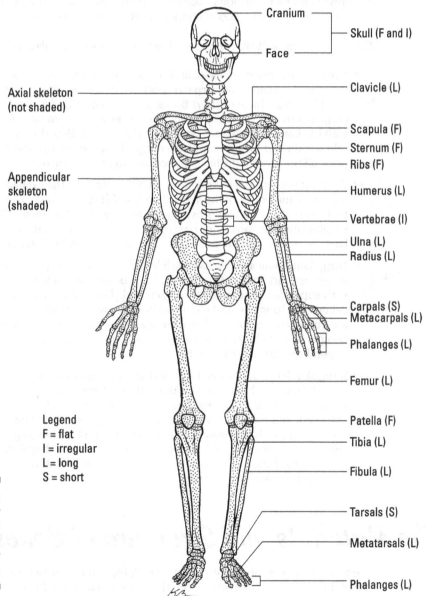

Cranium
Skull (F and I)
Face
Axial skeleton (not shaded)
Clavicle (L)
Scapula (F)
Sternum (F)
Ribs (F)
Appendicular skeleton (shaded)
Humerus (L)
Vertebrae (I)
Ulna (L)
Radius (L)
Carpals (S)
Metacarpals (L)
Phalanges (L)
Femur (L)
Legend
F = flat
I = irregular
L = long
S = short
Patella (F)
Tibia (L)
Fibula (L)
Tarsals (S)
Metatarsals (L)
Phalanges (L)

Figure 13-1:
The major bones of the axial and appendicular skeleton.

Bones aren't all the same shape. The shape of a bone affects, and reflects, its role in the skeleton. The bones of the legs are much bigger and stronger than the bones of the arms because bones of the lower extremity are loaded more during weight bearing and movement. Although there is a genetic blueprint to an individual bone, the ultimate shape and strength of a bone reflects its use and loading (from the pull of muscle and the support of body weight), as well as the individual's diet, hormonal factors, and other lifestyle factors. Bones adapt in shape to perform the roles of support, protection, and movement.

There are four classifications of bone based on shape (refer also to Figure 13-1):

- **Short:** Short bones are small and shaped like blocks. Short bones are found in the hands (the carpal bones) and in the feet (the tarsal bones). The block-like shape of short bones allows limited gliding motion. The carpals provide good support for the dexterous movements of the fingers with the hands in awkward positions. The tarsals allow the feet to adapt to odd terrain and to serve both as energy absorbers and as a solid lever to push off the ground during locomotion and other activities.

- **Flat:** Flat bones have flat surfaces. The flat bones of the axial skeleton include some bones of the skull, the ribs, and the sternum. Flat bones protect internal organs and provide sites for muscle attachment. The scapulae glide over the posterior part of the rib cage and provide a moveable base for muscle attachment, allowing the variety of motion of the arms.

- **Long:** Long bones serve as the rigid links of the body (see more on bones as rigid links in Chapter 8). Long bones have a long round shaft with expanded prominences at each end. The shaft of a long bone is called the *diaphysis,* and each end is called an *epiphysis.* Individual long bones have notable bony bumps at the site of muscle attachment. Long bones are located in the arms and legs, and in the hands and feet beyond the carpals and tarsals.

- **Irregular:** Irregular bones have odd shapes that don't fit into the other classifications. The irregular shape allows these bones to effectively perform multiple functions. The 26 bones of the vertebral column are a marvel of adaptation. These irregular bones protect the spinal cord; support the upper body against gravity when at rest and during movement; and allow movements of the upper body such as bending forward (flexion) and backward (extension), leaning to the side (lateral flexion), and twisting (rotation) of the trunk.

The Materials and Structure of Bones

The word *bone* describes both the physical, identifiable structure of the individual bones of the skeleton and the tissue that makes up the physical, identifiable structures commonly known as "a bone." A physical bone includes bone tissue and other tissues such as *marrow* (the fat-like substance that fills the interior of bones) and blood. Bones are richly served by blood vessels that provide the ability for bones to heal so well when damaged.

The tissues work together to allow a physical bone to serve the multiple roles in the skeletal system. In this section, I provide an overview bone tissue and then consider the physical structure of bone.

Materials: What bones are made of

Bone tissue consists of cells and an *extracellular matrix* (the structure holding the cells in place). In this section, I describe the basics of bone cells and the extracellular matrix.

There are three main types of cells in bone tissue:

✔ **Osteocytes:** Osteocytes are mature bone cells. They're firmly held in place in the extracellular matrix by a mineral coating. Osteocytes control the biological activity within a bone by responding to the mechanical loading imposed on bone.

✔ **Osteoblasts:** Osteoblasts are bone-building cells. The term *blast* is from the Greek word for "bud" — osteoblasts build new bone. An osteoblast lays down new extracellular matrix and then becomes trapped within the new matrix. Eventually, an osteoblast transforms into an osteocyte as the matrix hardens (mineralizes) around it, creating the hardness of bone. Osteoblasts create new bone as a bone grows, and they also create new bone to repair damage.

✔ **Osteoclasts:** Osteoclasts are the cells responsible for breaking down the existing extracellular matrix and releasing the minerals. The term *clast* is from the Greek word for "broken" — osteoclasts break down existing bone. This process is called *bone resorption.* The minerals released as bone is resorbed are transported from the site for use elsewhere in the body or to be passed from the body. Osteoclasts are active both to resorb existing bone as a bone grows and to remove damaged bone as part of the repair of damaged bone.

The extracellular matrix of bone consists of

✔ **A variety of fibers:** The main fiber in the extracellular matrix is collagen, a protein-based fiber. Collagen contributes to the tensile strength of bone, providing a lot of resistance to being stretched under the pulling load imposed by muscle tension. Collagen also gives bone *elasticity,* or the ability to return to its original shape after loading from the external forces imposed on bone as the body interacts with the environment.

✔ **A material called the ground substance:** The ground substance of the extracellular matrix of bone consists of many materials; the two most important materials mechanically are

- **Calcium:** Calcium makes up about 60 percent to 70 percent of a bone's weight. As a hard mineral, calcium provides most of a bone's resistance to deformation and fracture. Calcium is drawn to an osteoblast trapped in the extracellular matrix, and it adds strength to the bone as the calcium *crystallizes,* or hardens, within the matrix.

- **Water:** Water makes up about 25 percent to 30 percent of a bone's weight. Water contributes to the tensile and *compressive strength* (resistance to elongation and to being squished; see more on tension and compression in Chapter 12) of a bone.

Structure: How bones are organized

Within an individual bone, the bone tissue is arranged in two forms. The forms differ in *porosity,* the non-mineralized space within the bone tissue in the physical structure. In this section, I describe how the two arrangements of bone tissue create the general structure of a typical bone, and the role provided by each arrangement.

Cortical bone

Cortical bone (also known as *compact bone*) forms the hard outer shell, or *cortex,* of a bone (see Figure 13-2). Cortical bone is solid and dense with bone tissue. There is less than 5 percent to 30 percent of open space, or porosity, in compact bone. The non-mineralized space within cortical bone includes passages for nerves and blood vessels into the bone's interior.

The *diaphysis,* or shaft, of a long bone is shaped like a hollow tube. The hard walls of the tube consist of cortical bone. The hollow central region of the tube is called the *medullary cavity,* and it's filled with marrow and fat.

The bone cells and extracellular matrix are not randomly aligned in cortical bone. The bone tissue aligns through modeling and remodeling (explained later in this chapter) to provide greatest resistance in the typical directions of stress imposed on the bone. Cortical bone provides most of the stiffness or resistance to deformation when a compressive load is applied to the bone.

On the outer surface of cortical bone is a two-layer membrane called the *periosteum* (*peri* means "around," *osteum* refers to bone). On the inner surface of cortical bone is a similar membrane called the *endosteum* (*endo* means "within"). The outer layers of the periosteum and endosteum serve as protective sheets over the bone surfaces, while the inner layers help in bone modeling.

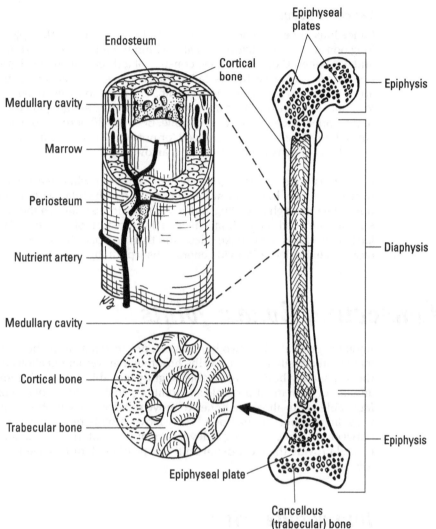

Figure 13-2:
Compact and trabecular bone.

Osteoclasts and osteoblasts are active on the surface of cortical bone under the periosteum and the endosteum. Dominant activity of the osteoblasts on the outer surface increases the circumference of the bone, while dominant activity of the osteoclasts on the inner surface keeps the width of the cortical bone shaft at an optimal thickness. I talk more about balancing the activity of osteoblasts and osteoclasts later in this chapter.

Cancellous bone

Cancellous bone is found within the interior of a bone, in the epiphyses of long bones (refer to Figure 13-2) and within the interior of short, irregular and flat bones. Cancellous bone consists of a three-dimensional gridwork of bony struts and cross-beams called *trabeculae*, giving cancellous bone the appearance of a sponge or honeycomb. The commonly used names of *cancellous* (meaning "lattice"), *spongy*, or *trabecular bone* are all good descriptors of the appearance of this form of bone. Between 30 percent and 90 percent of cancellous bone is non-mineralized, or porous. The non-mineralized space in cancellous bone is filled with bone marrow.

Cancellous bone provides great strength to a bone while keeping its overall weight relatively low to allow movement. The trabeculae within cancellous bone align through modeling and remodeling in response to the typical stresses imposed on the bone, maximizing strength of the bone. The trabeculae in cancellous bone make it less stiff than compact bone, so it deforms more when loaded to provide energy absorption by the bone.

Connecting Bones: Joints

A joint is present when two bones meet, or *articulate*, in the body. The three main joint classifications describe how much movement is allowed between the articulating bones: immovable, slightly movable, and freely movable. The amount of movement possible at a joint is called the *range of motion* (ROM). Moveable joints allow the bones to move relative to each other while also providing support to the body. Freely moving joints providing a large ROM between the bones serving as levers are of most interest in human movement. I provide the most detail on this type of joint, often called a *synovial joint*.

Immovable joints

An *immovable joint* allows no movement between the articulating bones. The joints between bones of the adult skull are immovable. The bones can move relative to each other at birth, but over time, the bones fuse at the joint to provide protection to the brain.

Slightly movable joints

Slightly movable joints allow limited movement between the bones. At a slightly movable joint, the bones are connected by a tough but flexible material called *fibrocartilage*. Fibrocartilage deforms as movement occurs at a slightly movable joint. The intervertebral disks between individual vertebrae of the vertebral column are slightly movable joints. The disks provide support for the upper body and allow a limited range of motion between adjacent vertebrae.

Freely movable joints

Freely movable joints (also known as *synovial joints*), as expected from the name, allow the most range of motion of any type of joint. All joints of the appendicular skeleton, from the hips to the toes and from the shoulders to the fingers, are synovial joints.

Synovial joints vary in structure and by how great a range of motion is possible at the joint. However, synovial joints include the common features shown in Figure 13-3.

Two bones meet. Of course they do, that's why it's a joint. A joint capsule encircles a synovial joint, creating a self-contained "bag" around the ends of both bones. The joint capsule has two layers:

- ✔ **An outer layer of a flexible fibrous material:** The outer layer holds the bones together and provides support for the joint. The fibrous material of the joint capsule joins into the periosteum of each of the bones.

- ✔ **An inner layer called the *synovial membrane* (hence the name synovial joint):** The synovial membrane secretes a fluid called (you may have seen this coming) *synovial fluid.* The interior of a synovial joint (sometimes called the *synovial cavity* — imagine that) is filled with synovial fluid. The synovial cavity is actually quite small because the joint capsule fits snugly around the joint.

Synovial comes from the Greek word *syn* meaning "with" and the Latin word *ovum* meaning egg — an apt descriptive name because synovial fluid has the appearance and consistency of egg white.

Articular cartilage covers the end of each bone. The layer of articular cartilage increases the surface area of the articulating surfaces to spread the forces at the joint over a wider area and reduces the stress on the cartilage. The most important role of cartilage in movement is to reduce the shear or friction between the articulating bones. Wet articular cartilage against wet articular cartilage has a very low coefficient of friction (see Chapter 4 for more on the coefficient of friction), creating almost no friction and allowing easier sliding of the bones across each other during movement.

Articular cartilage is wet because it acts like a sponge for synovial fluid as the joint moves through its range of motion. Only part of the articular cartilage at the end of each bone contacts the cartilage of the other bone at any joint position. As the bones move, synovial fluid is squeezed out of the articular cartilage where the ends are in contact, and synovial fluid is sucked in to the articular cartilage where the ends are not in contact. The synovial fluid is the only source of nourishment and maintenance for articular cartilage, and the sponge-like action created by loading the cartilage during movement is critical to maintain a healthy joint.

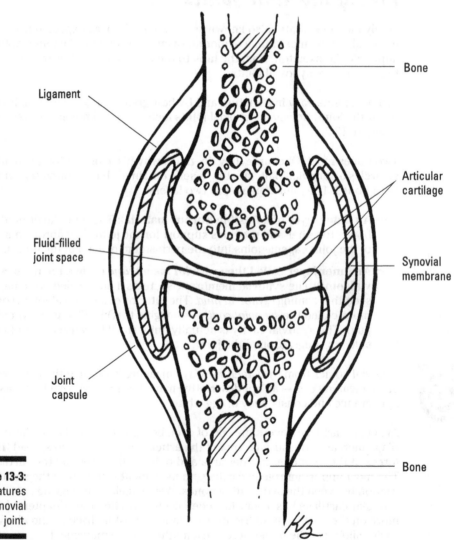

Ligament

Bone

Articular cartilage

Fluid-filled joint space

Synovial membrane

Joint capsule

Bone

Figure 13-3: The features of a synovial joint.

Articular cartilage has limited or no blood supply. Because blood supply is important for healing a damaged tissue, cartilage can't heal if injured. The degeneration of articular cartilage is known as degenerative joint disease, or osteoarthritis. I explain the basics of osteoarthritis in more detail later in this chapter.

A ligament joins bone to bone and reinforces the joint capsule. Ligament consists of collagen fibers and elastin fibers, and these fibers stretch to keep the joint stable as the bones move. Some ligaments are simply overgrown areas of the joint capsule, but others are specialized structures within the joint. Damage to a ligament is called a *sprain*. A sprain occurs when joint loading stresses a ligament beyond its elastic limit (see Chapter 12 for more on the elastic limit).

All ligament sprains are not the same. There are three grades of sprain, reflecting how much the ligament gets stretched beyond its elastic limit:

- ✔ **Mild sprain (I):** The joint is painful to move and tender to touch. The elastic limit of the ligament was barely exceeded.

- ✔ **Moderate (II):** There is a lot of swelling, and the joint feels loose. The loading stresses the ligament close to its ultimate strength.

- ✔ **Severe (III):** The failure point of the ligament was exceeded, and the fibers are completely disrupted, or torn. The joint is completely unstable because the ligament can't provide support.

Despite the name, no joint is exactly a "freely movable joint." The ROM is limited by the shape of the ends of the bones that meet at the joint, by restriction from the joint capsule and ligaments, by other soft tissue including muscle and fat on the segments that meet, by temperature, and by past injury. For more on the ROM at specific joints, and for more details on the variety of synovial joints, pick up a copy of *Anatomy & Physiology For Dummies*.

Growing and Changing Bone

Living bone is a dynamic structure. Bone continually changes, from early development of the fetus until death. A newborn's skeleton is mostly cartilage, but as the child grows, the cartilage *ossifies,* or transforms into hard bone. Bones obviously get longer as the child gets taller, but they also increase in width to support a greater body weight and stronger muscle forces. Changes in bone length are typically complete by 18 to 25 years of age, but bone width continues to vary throughout life.

In this section, I explain how bone dimensions change and the role of mechanical stress on bone changes.

Changing bone dimensions

Three main processes underlie the changes in bone length and width:

- ✔ **Ossification:** Ossification is the process of transforming cartilage into bone. The majority of ossification occurs during childhood, as bone length and width change most rapidly, but ossification also occurs during the early phase of healing a bone fracture.

- ✔ **Modeling:** Modeling is the process of adapting the overall shape of a growing bone in response to the stresses from loading. Modeling refines the dimensions of a bone during growth, and it also occurs during the final phases of fracture healing. Modeling typically creates a stronger bone and increases bone mineral density (the amount of bone tissue in a given space of bone).

- ✔ **Remodeling:** Remodeling is the process of adapting the structure of an existing bone at a particular site in response to an altered level of mechanical loading. Remodeling can increase bone mineral density and bone strength or, unfortunately, decrease bone mineral density and bone strength, as I explain later.

Growing longer

At birth, most bones of the human skeleton are simply a cartilage model of the eventual shape. The central part of the *diaphysis,* or shaft of the bone, is typically ossified at birth, and soon after birth, the *epiphysis,* or ends of the bone, also ossify. A plate of cartilage remains between the ossified diaphysis and epiphysis. Each plate is called an *epiphyseal plate,* or simply a growth plate. Figure 13-2 shows the location of the epiphyseal plates in a typical long bone. When a bone grows longer, this is called *longitudinal growth.*

Although all long bones have an epiphyseal plate, one is not always present at both ends of all bones.

Bone lengthening occurs at an active epiphyseal plate. Essentially, new cartilage is laid down by the epiphyseal plate, lengthening the bone. The cartilage closest to the diaphysis then ossifies. As ossification occurs on the diaphysis side of the plate, more new cartilage is laid down on the epiphyseal side of the plate. The process continues until the epiphyseal plates close and the diaphysis and epiphysis fuse.

The epiphyseal plates typically close, ending long growth of the bone, by the end of adolescence (about 18 years of age), but some don't close until early adulthood (about 25 years of age). About 90 percent of a bone's mineral content is deposited by the time the epiphyseal plates close, and the peak bone mineral content of a bone should not be reached before an individual is 20 to 30 years old.

Even as the bone lengthens at the epiphyseal plate, bone modeling occurs. The activity of the osteoclasts and osteoblasts refines the shape of the newly produced bone in response to the mechanical stresses imposed by body weight and muscle forces.

Growing wider

Bones change in width, or diameter, throughout a lifetime. This is called *appositional growth.* The changes in width are most rapid before the epiphyseal plates close, as part of the bone modeling process to keep bone width proportional to the increasing bone length. The bone modeling adapts the bone to withstand the greater stress from the forces incurred as the more massive growing body is supported and moved. After plate closure and continuing to the end of life, bone remodeling occurs in response to the mechanical stress imposed on the bone. Remodeling can cause an increase or decrease in the width of the bone.

The changes in bone width during modeling and remodeling reflect the net activity of the osteoclasts and osteoblasts working on the inner and outer surfaces of the cortex of the bone. The changes in bone width are reflected in a changing diameter of the entire bone and a changing thickness of the *bone cortex,* or hard outer shell.

Osteoblasts build new bone and osteoclasts remove old bone.

On the outer surface of the cortex, the activity of the osteoblasts dominates over the activity of the osteoclasts. The bone grows wider because the amount of new bone added is greater than the amount of bone resorbed. On the inner surface of the cortex, the activity of the osteoclasts dominates over the activity of the osteoblasts. The medullary cavity gets wider. Overall, the bone gets wider and the cortex adopts an ideal thickness to maintain a light but strong bone.

Stressing bone: The effects of physical activity and inactivity

Heredity, or genetics, is a dominant factor in the shape and strength of bones. The blueprint for the shape of the skeleton is part of human heredity. But differences exist between individuals in the strength of bones because the quality and alignment of the bone tissue varies. Bone strength differences reflect heredity, but other factors — including diet, disease, previous injury, lifestyle choices (including alcohol and tobacco use), and physical activity — are known to affect the quality and alignment of bone tissue.

Physical activity is of particular interest in biomechanics because the forces during activity create stress and strain within the bones. (I explain stress and strain in Chapter 12.) Bone deformation stimulates the osteocytes to activate the bone modeling and bone remodeling activity of osteoblasts and osteoclasts.

In this section, I outline the role of mechanical stress on the quality and alignment of the bone tissue produced during modeling and remodeling. Because the basic processes of bone modeling and bone remodeling are the same, I'll use the term *adaptation* to include them both.

Outlining the effect of stress on a bone: Wolff's law

Wolff's law states that a bone adapts to the magnitude and direction of the stress it receives over a period of time. Wolff's law means the following:

- ✔ **With an increase in stress, more bone tissue is added, with a higher mineral content that's aligned in the direction of the stress.** This adaptation is called *hypertrophy* (*hyper* means "more," and *trophic* means "growth"). As bone hypertrophies, the bone mineral density (BMD) increases.

- ✔ **With a decrease in stress, the amount of bone tissue is reduced and the mineral content in the bone that remains is reduced as well.** This adaptation is called *atrophy* (*a* means "without" and *trophic* means "growth"). As bone atrophies, the BMD decreases.

Figure 13-4 shows the effect of changing stress on the adaptation of *cortical bone*, as represented in the diaphysis of a long bone. With hypertrophy, the width and thickness of the cortex increase, and BMD increases. With atrophy, the width and thickness of the cortex decrease, and BMD decreases.

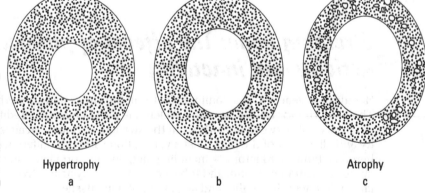

Figure 13-4: Adaptations in cortical bone to the level of stress.

Hypertrophy
a

b

Atrophy
c

Figure 13-5 shows the adaptation of *cancellous bone* to changing stress. With hypertrophy, the *trabeculae* (the shafts of bone making up the lattice of cancellous bone) grow thicker and more numerous. With atrophy, the trabeculae become thinner as bone content decreases. In extreme atrophy, as shown in Figure 13-5, the integrity of the lattice is reduced as some trabeculae are lost and those that remain are thinner and contain less bone tissue. This represents a condition called *osteoporosis,* which I explain later in this chapter.

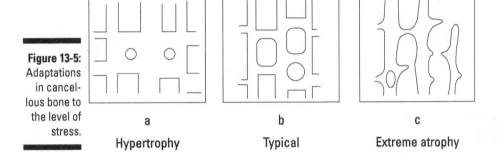

Figure 13-5:
Adaptations in cancellous bone to the level of stress.

a	b	c
Hypertrophy	Typical	Extreme atrophy

Figure 13-6 shows the differences in the stress-strain curves for bone with different levels of BMD (see Chapter 12 for more on the stress-strain curve). The stress-strain curve for bone with the optimal BMD shows a linear elastic region and a short plastic region after the elastic limit (EL) before the failure point. With high BMD, the linear elastic region shows greater stiffness, or less strain (deformation) for an applied stress, a higher elastic limit, and a higher failure point compared to the optimal BMD. However, the short plastic region means the bone fails soon after reaching its EL. With low BMD, the linear elastic region shows reduced stiffness, or more strain for an applied stress, a lower elastic limit, and a lower failure point compared to the typical bone.

Bone is an anisotropic material because it has a different strain response depending on the direction of the applied stress. Bone strength is highest against compressive loading (about 170 Mpa) and lowest against shear loading (about 52 Mpa). Tensile strength falls in between, at about 110 Mpa. These values mean bone best resists pushing forces (created in supporting the body and maintaining alignment of segments), least resists torsion (created as bone is twisted around its long axis), and does well against pulling forces (created as a tendon transmits muscle force to the bone).

The adaptation of bone tissue reflects the summed activity of the osteoclasts and the osteoblasts. If osteoclast activity dominates, the bone atrophies. If osteoblast activity dominates, the bone hypertrophies. Although many factors affect bone adaptation, three lifestyle factors are critically important:

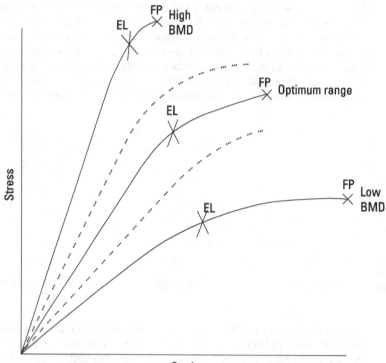

Figure 13-6:
Stress-
strain
curves for
bone.

✔ **Calcium availability:** Calcium provides strength to bone, and it's also important in the activity of muscles and nerves. Without adequate calcium intake, bone resorption releases calcium for use by muscles and nerves, but this decreases bone strength. Calcium serves as a "building block" of bone and must be available to maintain and build bone strength. Along with calcium, Vitamin D is important because this vitamin helps the body take up calcium from the food, drink, or supplement consumed to be used by bones, muscles, and nerves.

✔ **Physical activity:** Physical activity provides a stimulus to increase bone strength through the bone adaptation process. The forces applied to bones during physical activity strain the osteocytes and bone tissues and actually cause minor structural damage within the tissue. On signals from the strained osteocytes, osteoclasts remove bone at the damaged sites and osteoblasts lay down new and stronger bone, in accordance with Wolff's law. A variety of physical activity involving the upper and lower body helps to impose stress on all the bones of the body. The activity should be vigorous, involving multiple muscle groups. Walking is more stimulating than standing. Stair climbing is more stimulating than walking. Jumping up and down is more stimulating than stair climbing. Strength-training exercises using the arms and trunk are very beneficial for the upper body. Of course, the adaptation of bone takes time, so adequate rest should be provided between similar activities. The general rule is to take 48 hours between sessions of a similar activity.

✔ **Hormonal regulation:** Of special importance are the hormones estrogen and testosterone. These hormones are produced in the reproductive organs, estrogen in the ovaries of women and testosterone in the testes of men. Among other roles in the body, these two hormones regulate osteoclast activity. Men luck out because testosterone production is not as affected by getting older as is estrogen production in women. Estrogen production is related to menstruation. When menstruation ends at menopause, estrogen production declines a lot. Irregular menstruation before menopause can also cause estrogen levels to drop. With a decreased level of estrogen, the osteoclasts are less inhibited and the rate of bone resorption increases leading to a decrease in bone strength. Because of its important role in regulating osteoclast activity, women should consult a physician if menstruation is irregular or has ended.

Of the many factors affecting bone health, genetics dominates. Nothing can change the genetics of an individual. However, the three lifestyle factors of diet, exercise, and hormonal regulation are choices, and good choices can beneficially influence the expression of the genetic potential of bone health.

Calcium availability, activity, and hormonal levels are linked in maintaining bone health, and making lifestyle choices to increase bone strength must consider all three. More calcium in the diet increases the availability of the building block of bone, but it won't be as effective without the bone-building stimulus of physical activity or if osteoclast activity is not inhibited. Similarly, becoming more physically active adds the stimulus for bone hypertrophy, but more activity alone will not increase bone strength without adequate calcium available to serve as the building block for bone tissue.

Too much bone hypertrophy: Osteoarthritis

Osteoarthritis is a progressive disease of a synovial joint. Osteoarthritisis is also called degenerative joint disease because, as shown in Figure 13-7, the articular cartilage gets thinner and the ends of the articulating bones become misshapen with *osteophytes,* or bone spurs. In advanced cases of osteoarthritis, there is bone-on-bone contact.

Osteoarthritis is a painful condition that interferes with the use of the joint. About 50 percent of people who get osteoarthritis show symptoms of the disease by age 65, and the emotional costs associated with the loss of joint use and the financial costs of pain medication and medical treatment are high. For many with advanced osteoarthritis in the hip or knee, restricted mobility and increased pain lead to joint replacement surgery, involving insertion of an artificial, or prosthetic, joint.

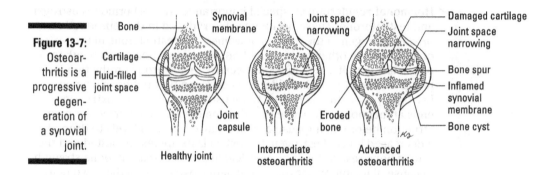

Figure 13-7: Osteoar-thritis is a progressive degen-eration of a synovial joint.

Osteoarthritis was long thought to be a disease of the elderly — old cartilage simply wore out. But research suggests a biomechanical basis for the dis-ease, beginning many years before the symptoms show up. One idea is that abnormally high mechanical stress at the joint hypertrophies the ends of the bones, leading to greater stress on the cartilage and causing its gradual degeneration over many years. The degenerating cartilage further alters the stress pattern on the bones and cartilage, and the disease progresses with its negative consequences.

Mechanically related risk factors for osteoarthritis include the following:

✔ **Genetics:** Men tend to get more osteoarthritis in the hips, and women tend to get more it in the knees. Because women tend to have wider hips than men, the differences in osteoarthritis may be related to the differ-ences between women and men in the stress distribution at these joints. Also, a family history of osteoarthritis at a particular joint might reflect an inherited predisposition to the disease.

✔ **Body weight:** Heavier individuals are more likely to develop osteoarthri-tis in the lower extremities than lighter individuals, because stress on the hip and knee joints increases with higher body weight.

✔ **Previous trauma:** Suffering a serious joint injury increases the risk of developing osteoarthritis in the affected joint. Damage to the cartilage at the initial injury, increased *laxity* (looseness) of the joint from a ligament sprain, and/or direct trauma to the bones when the joint was injured can alter the stress pattern at the joint.

A wolf in sheep's clothing

In the late 1960s and early 1970s, Dr. Eric Radin and his colleagues first proposed the mechanical basis for the development of osteoarthritis, based on the biomechanics of bone and cartilage, records of osteoarthritis among different workers (identifying more frequent and serious development of the disease in heavily loaded joints), and research measuring joint loading.

To test their idea that high stress caused osteoarthritis, they used a flock of sheep. Half the sheep were kept in a grassy pasture, and half were kept in a corral with a concrete floor. The leg joints of the sheep kept on the concrete floor showed greater joint degeneration after a shorter period of time than the sheep kept on the grassy pasture.

Too much atrophy: Osteoporosis

Osteoporosis, literally meaning "porous bone," is a progressive disease. Before osteoporosis there is *osteopenia,* or low BMD. As bone atrophy continues, the BMD gets extremely low in both compact and cancellous bone, the lattice of trabeculae in cancellous bone thins and may disappear (as described earlier), and bone strength, or fracture resistance, becomes extremely low. An osteoporotic fracture is a fracture from a load easily withstood by healthy bone. Common sites of osteoporotic fractures are just above the wrist in the radius, the proximal end of the femur (called a broken hip), and the thoracic vertebrae at the level of the shoulder blades. The first two are associated with a fall, but the vertebrae fractures occur from supporting the upper body.

Osteoporosis is most common in elderly women because a reduced estrogen level following menopause increases the rate of BMD decrease. Reversing osteoporosis and osteopenia is very difficult, so preventing BMD from reaching low levels is important. Three factors related to keeping BMD high are important and somewhat under the control of an individual because they are affected by diet, activity, and maintaining hormone levels (as described in the preceding section):

- **Reaching a high peak BMD:** Losing BMD is inevitable because of genetic factors, so reaching a high BMD before it begins to decline is important. The best time to increase BMD is when the bones are rapidly growing, especially the teenage years.

- **Reaching the peak BMD at an older age:** Unfortunately, many people reach peak BMD in the late teens or early 20s even though all the epiphyseal plates aren't closed until later.

✔ **Reducing the rate of loss of BMD:** After reaching peak BMD, continuing with lifestyle choices that stimulate bone modeling can reduce the rate at which BMD decreases.

By reaching a higher bone density at a later age, and then reducing the rate at which BMD is lost, an individual can hope to delay the risk of osteopenia and osteoporosis. This means that proper diet, exercise, and hormonal regulation are important throughout the life span. There are no "critical" years when BMD is of any more concern than at any other time; developing and maintaining healthy bones takes a lifetime. Prevention is more effective than treatment.

Chapter 14

Touching a Nerve: Neural Considerations in Biomechanics

In This Chapter

▶ Understanding what the nervous system does

▶ Looking at the central and peripheral nervous systems

▶ Typecasting neurons

▶ Sensing muscle length and muscle tension

▶ Activating motor units

*T*he nervous system is the body's detecting, communicating, and controlling system. The human nervous system is probably the most complex system in the universe. How it works is the focus of intense research in a variety of different fields of science.

The musculoskeletal system is controlled by the nervous system. The nervous system collects information about the body and its environment using sensory receptors, communicates the information using electrical and chemical signals, and controls the muscle response.

In this chapter, I focus on the nervous system control and monitoring of muscles. First, I provide a basic description of the major components of the nervous system, the central and peripheral nervous systems. Then I describe the neuron, because it's the basic unit of the nervous system, and provide details on three types of neurons important to controlling and monitoring the muscle system. I describe how neurons and two important sensory receptors are arranged to control the muscles and provide an overview of how the nervous system controls the tension, or force, produced by a muscle.

Monitoring and Controlling the Body: The Roles of the Nervous System

The nervous system has several key roles in the body:

- **Collecting sensory input:** The nervous system collects information from inside and outside the body using sensory receptors. These receptors inform the body about the environment around it and the internal status of the individual systems that make up the body.

   Sensory receptors include the eyes, ears, nose, taste buds, and structures called *proprioceptors,* which are specialized to detect other information such as touch, temperature, and pain. Proprioceptors also include two specialized structures located in the skeletal muscles of the body. These specialized structures detect the length of a muscle and the amount of muscle tension, or force, it's producing. I describe these two sensory receptors — the muscle spindle and tendon organ — later in this chapter.

- **Integrating sensory input and initiating the motor output:** The information received from the sensory receptors is transmitted to the central nervous system for processing. Any muscle response, or motor output, to the sensory input is initiated within the central nervous system. (I describe the central nervous system in more detail a bit later.)

- **Transmitting motor output:** The motor response initiated in the central nervous system is sent out to control the tension or force produced by the muscles. This response is called *motor output* because the muscles act as "motors" to control the bones of the body.

The detection of sensory information and the transmission of the motor output response occur through the peripheral nervous system, all the neural tissue outside the central nervous system.

Outlining the Nervous System

The nervous system contains billions of individual nerve cells called *neurons.* The neuron is the basic unit of the nervous system. The interface between two neurons is called a *synapse.* The neurons and synapses create a widespread communication and control system within the body.

The communication signal along a neuron is a small electrical impulse of about 100 millivolts (one-tenth of a volt) called an *action potential.* A neuron can generate an action potential in response to the input it receives from other neurons at the synapses. The input from another neuron consists of a chemical substance called a *neurotransmitter.* Basically, one end of a neuron is specialized to receive the neurotransmitter from other neurons, and the other end is specialized to release the neurotransmitter to other neurons. (I describe the neuron in more detail later in this chapter.)

The nervous system can be divided into two main parts: the central nervous system and the peripheral nervous system. The general outline of the nervous system is shown in Figure 14-1.

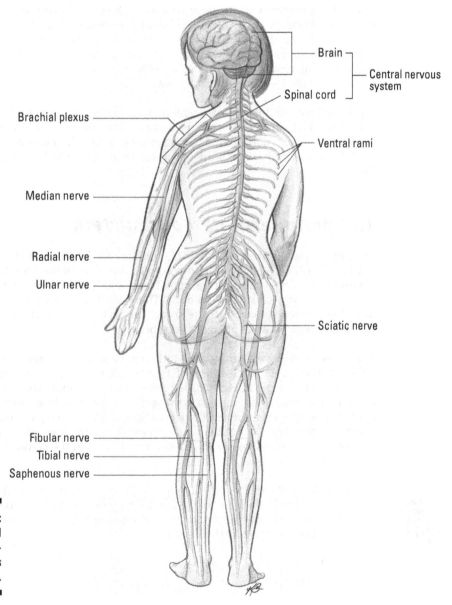

Figure 14-1: The central and peripheral nervous systems.

The central nervous system

The central nervous system (CNS) is the largest part of the nervous system, consisting of the brain and the spinal cord. The brain is contained within the head, where it's protected by the bones of the skull. The spinal cord comes off the brain and runs down the midline of the back, contained within and protected by the *vertebrae,* the individual bones of the spinal column.

The brain is the site of thinking and knowing. Sensory input is perceived in the brain; thoughts are created here; and memories, including patterns of physical activity, are stored here. Any purposeful movement — such as walking or running, reaching or grasping, throwing or catching — is initiated in the brain. The motor output to perform the activity is sent down the spinal cord to activate the necessary muscles.

The spinal cord is the major transmission line of the nervous system. All action potentials related to sensory input and motor output travel along the spinal cord. Some motor response to sensory input is initiated in the spinal cord. I describe this spinal cord–generated response, called a *reflex,* later in this chapter.

The peripheral nervous system

The peripheral nervous system (PNS) contains all the nervous system outside the CNS. The nerves are the dominant feature of the PNS (refer to Figure 14-1). Nerves come off the spinal cord and brain in pairs, with one branch going to each side of the body. There are 31 pairs of spinal nerves branching off the spinal cord. The spinal nerves leave the vertebral column through spaces between adjacent vertebrae.

A *nerve* is a bundle of neurons. The individual neurons in a nerve transmit either sensory input toward the CNS (called *afferent input*) or motor information toward the target muscle (called *efferent output*). Both types of neurons are bundled within a nerve. As a nerve gets farther away from the spinal cord, it branches even more. The nerves aren't protected by any bones. Damage to the spinal cord affects the nerves below the level of damage, resulting in a loss of sensory and motor function when the communication to the brain is lost.

Together, the CNS and PNS form a loop system. Sensory input, or afferent information, travels along the nerves of the PNS toward the CNS from around the body, and the motor output, or efferent information, travels away from the CNS around the body along the nerves of the PNS.

Zeroing In on Neurons

The neuron, or nerve cell, is the basic unit of the nervous system. The parts of a neuron allow it to receive input from and transmit output to other neurons or to muscles.

Parts of neurons

Each neuron consists of three basic parts:

- **A cell body:** The body of a neuron contains structures that direct its activity, including metabolism and growth.
- **Dendrites:** The dendrites are thin projections from one end of the cell body. Dendrites serve as the main receptor sites of a neuron. The dendrites receive information in the form of the neurotransmitter from another neuron. In response to the neurotransmitter, a small electrical impulse called a *graded potential* is sent along the dendrite to the cell body.
- **Axon:** The axon is a long projection from the end of the cell body opposite to the dendrites. At its end, the axon divides into thin branches called *collaterals*. If an action potential is generated by a neuron in response to the neurotransmitter input it receives from other neurons, the action potential travels away from the cell body along the axon and goes out along each collateral branch. Each collateral branch of the axon synapses with either another neuron or a muscle fiber.

Types of neurons

Neurons come in three basic types: motor neurons, interneurons, and sensory motor neurons. Motor neurons send information from the CNS to the PNS. Interneurons transmit information between neurons within the CNS. Sensory neurons send information to the CNS from the PNS. I describe each type of neuron in more detail in the following sections.

Motor neurons

A *motor neuron,* or afferent neuron, transmits an action potential from the CNS to a destination outside the spinal cord. Some motor neurons control the glands and organs of the body. Of most interest in biomechanics are the motor neurons transmitting action potentials from the CNS to the skeletal muscles. First, I describe the structure of a motor neuron; then I describe the organization of a motor unit.

The structure of a motor neuron

The dendrites and cell body of a motor neuron are located in the spinal cord, where they synapse with interneurons. These interneurons bring action potentials to the motor neuron from the brain, to initiate and control purposeful movement, or from a sensory neuron of the PNS, providing input on the status of the muscle system. A motor neuron can be activated by either conscious (from the brain) or unconscious (from a sensory neuron) stimulation.

The *axon hillock* is the most excitable part of the motor neuron. It's located where the axon projects from the cell body. All the graded potentials from the dendrites of the motor neuron go toward the axon hillock. If enough graded potentials reach the axon hillock at the same time, an action potential is generated to travel out and along the axon of the motor neuron, and then to travel along each of the collaterals.

The axon of a motor neuron can be up to 1 meter long, because some go from the spinal cord to the muscles of the hand or foot. To allow the action potential to travel faster, a fatty substance called *myelin* wraps around the axon to create an insulating coating. The action potential jumps along the axon between gaps in the myelin called the *nodes of Ranvier.* Action potentials can travel along a myelinated axon at up to 140 meters per second.

The motor unit

The *motor unit* is the fundamental unit of the neuromuscular system for the control of the muscle. A motor unit consists of a single motor neuron and all the muscle fibers it innervates through its collaterals, as shown in Figure 14-2. The motor neuron in a motor unit is called an α (alpha) motor neuron. All the muscle fibers of a motor unit are located within the same muscle, although the fibers are spread throughout the muscle belly. (I cover muscle fibers and tension development in Chapter 15.)

When an action potential is generated in the axon of a motor unit, the action potential travels along each of the collaterals to synapse with each of the fibers. The synapse between a motor unit and a fiber is called the *motor end plate.* At the synapse, the action potential causes the release of a chemical transmitter that flows across the motor end plate to initiate an action potential on the muscle fiber. The fibers in the motor unit develop tension equally in an all-or-nothing event. The motor unit creates a subset of the fibers in a muscle that can be activated or deactivated as a single unit, allowing one way to control the total force produced by the muscle.

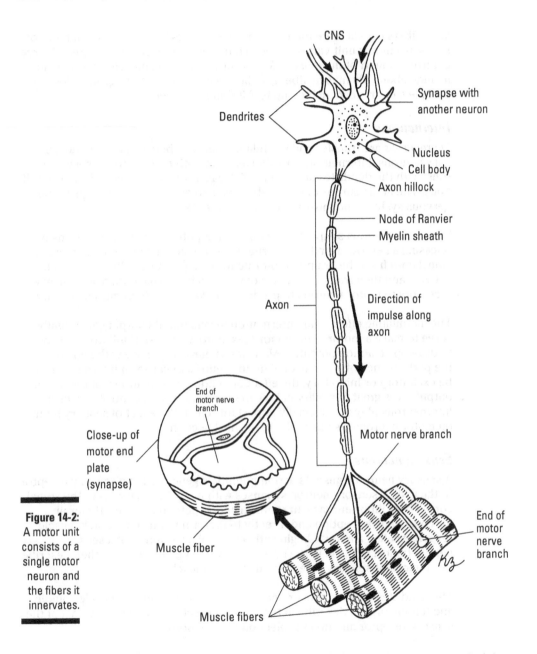

Figure 14-2:
A motor unit consists of a single motor neuron and the fibers it innervates.

Most muscles have several hundred motor units, allowing for control of the force produced. The number of fibers in a motor unit determines the amount of tension the motor unit produces when an action potential is generated.

More fibers equals more force; fewer fibers equals less force. The number of fibers in a motor unit varies between muscles and reflects the degree of force control required of the muscle. Motor units activating muscles of the hand include about 100 muscle fibers, while the motor units to large muscles in the back, leg, or arm can include up to 2,000 muscle fibers.

Interneurons

Interneurons are transfer and modulating neurons between motor and sensory neurons. Interneurons are shorter than both motor and sensory neurons, with the dendrites and axon collaterals projecting directly from the cell body. The interneurons make up about 99 percent of all the neurons in the nervous system, and most are located in the CNS.

Some interneurons simply transfer an action potential between neurons by releasing a neurotransmitter. Interneurons transfer signals coming down the spinal cord from the brain to motor neurons to produce voluntary muscle activity, and they transfer signals coming into the spinal cord from sensory receptors to motor neurons to produce *involuntary* (reflex) muscle activity.

The modulating effect of an interneuron means that its output can actually serve to make a motor neuron more (excitatory) or less (inhibitory) likely to develop an action potential. Many interneurons synapse with and affect the performance of an individual motor neuron. Because an interneuron can be excitatory or inhibitory, the effect of multiple interneurons on the motor output gives great flexibility in how the nervous system controls a muscle. Interneurons play an especially important role in the effect of sensory input on motor neurons, as I describe in the next section.

Sensory neurons

A sensory neuron transmits action potentials generated at a sensory receptor to the CNS. A sensory neuron synapses with many interneurons in the spinal cord. These interneurons can route an action potential toward the brain for conscious perception and directly to motor neurons for a quick muscle response called a *reflex*. I begin with a general description of a sensory neuron typical of those in the PNS, as shown in Figure 14-3, and then describe two important sensory receptors in skeletal muscle.

The dendrites of some sensory neurons, like those from the muscle spindle and tendon organ I describe in the following sections, are actually part of the sensory receptor and don't project off the cell body.

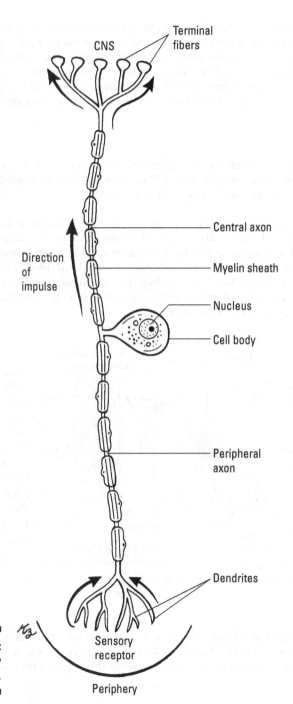

CNS

Terminal fibers

Central axon

Myelin sheath

Direction of impulse

Nucleus

Cell body

Peripheral axon

Dendrites

Sensory receptor

Periphery

Figure 14-3: A sensory neuron.

The cell body of a sensory neuron is located away from the dendrites, close to, but just outside, the spinal cord. The cell body projects off the axon, which runs from the dendrites at the sensory receptor to the spinal cord. The action potential transmitted along a sensory neuron is generated by the sensory receptor, a specialized structure sensitive to some sort of stimulus. In the spinal cord, the axon branches and these collaterals synapse with interneurons.

Muscle spindle

The *muscle spindle* is a sensory receptor located within the belly of a muscle, and it lies parallel to the muscle fibers. It detects the change in length of the muscle and the rate of change in length of the muscle.

The muscle spindle consists of a capsule of connective tissue enclosing 3 to 12 miniature skeletal muscle fibers called intrafusal fibers (see Figure 14-4). The intrafusal fibers are attached to the muscle spindle capsule which itself is attached to the connective tissue around the *extrafusal fibers* (the larger skeletal muscle fibers of the muscle belly). An individual muscle has many spindles within its fibers, distributed throughout the belly of the muscle and attached to fibers from different motor units.

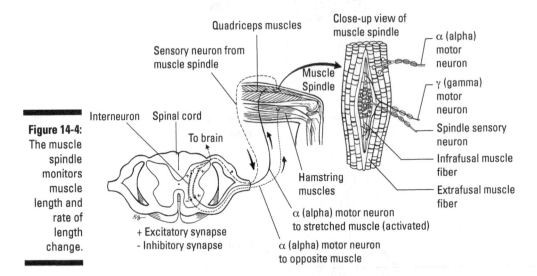

Figure 14-4: The muscle spindle monitors muscle length and rate of length change.

A motor neuron called a γ (gamma) motor neuron innervates the intrafusal fibers of the muscle spindle. The dendrites of a sensory neuron wrap around the intrafusal fibers within the capsule to create the sensory receptor.

When the muscle is stretched, the extrafusal fibers in the muscle spindle lengthen, deforming the sensory neuron dendrites wrapped around them. The deformation generates an action potential in the sensory neuron. Both the length of the stretch (the *static response*) and the rate of stretching (the *dynamic response*) affect the magnitude of the action potential generated. The action potential is transmitted along the sensory neuron back to the spinal cord where the axon branches to synapse with and generate action potentials in excitatory and inhibitory interneurons, as shown in Figure 14-4.

The excitatory interneuron synapses with the α motor neurons to the muscle being stretched. If the dynamic response stimulus from the sensory neuron is great enough, an action potential is generated in the α motor neurons and the rapidly stretching muscle is activated eccentrically — the muscle develops tension while it's lengthening. (Eccentric activity of a muscle is described in Chapter 15.) This response is called the stretch reflex — the sensory receptor serves to activate the motor neurons in the muscle being stretched. The faster the rate of stretch, the greater the activation of the muscle during the stretch reflex.

The inhibitory interneuron synapses with the α motor neurons to the muscle on the opposite side of the joint to the muscle being stretched. This inhibits the activation of the muscle. This response is called *reciprocal inhibition* — the sensory receptor serves to inhibit, or prevent activation of, the motor neurons in the muscle that could cause additional lengthening of the muscle being stretched.

If the muscle is stretched slowly, or if the muscle is stretched and held at the longer length, the action potentials from the sensory neuron don't cause the stretch reflex. Instead, an action potential is generated in the γ motor neurons innervating the intrafusal fibers. These fibers then shorten back to their original length in the capsule, essentially resetting the muscle spindle to respond to a stretch that begins from the new length.

The muscle spindle and stretch reflex are important contributors to the stretch-shorten cycle of optimal muscle tension development, as described in Chapter 15.

Tendon organ

The tendon organ is a sensory receptor located where the muscle fibers meet with the fibers of the tendon attaching the muscle to bone. Tendon organs are present at any muscle-tendon junction. The tendon organ detects the *magnitude,* or amount, of tension produced by the muscle.

As shown in Figure 14-5, the tendon organ consists of a capsule of connective tissue enclosing the dendrites of a sensory neuron. The dendrites intertwine among the fibers of the tendon and the connective tissue around the muscle fibers to create the sensory receptor. There are multiple tendon organs at each attachment of a muscle, and each tendon organ includes fibers from multiple motor units.

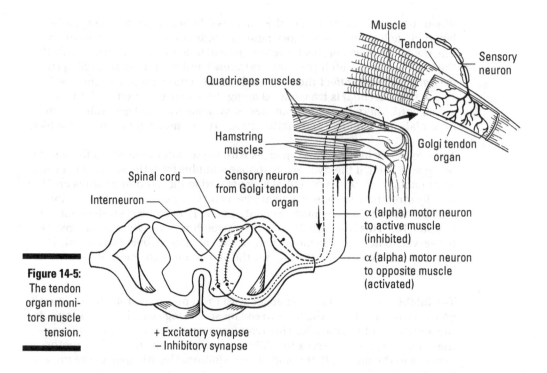

When the muscle develops tension, the muscle fibers and the fibers of the tendon are stretched, deforming the dendrites in the tendon organ. The deformation generates an action potential in the sensory neuron. More tension produces more deformation of the dendrites. The action potential generated is greater with more dendrite deformation, so the tendon organ output is proportional to the amount of tension produced by the muscle.

The action potential is transmitted along the sensory neuron back to the spinal cord where the axon branches to synapse with and generate action potentials in excitatory and inhibitory interneurons, as shown in Figure 14-5.

The inhibitory interneuron synapses with the α motor neurons to the muscle developing tension. The inhibitory effect of the interneuron reduces the activation of the motor units in the muscle developing tension. This response is called the *inverse stretch reflex* — the tendon organ serves to deactivate the motor neurons in the muscle developing tension. The higher the amount of tension produced in the muscle, the greater the deactivation of the muscle during the inverse stretch reflex.

The excitatory interneuron synapses with the α motor neurons to the muscle on the opposite side of the joint to the muscle developing tension. This activates the opposing muscle to resist the joint action caused by the muscle originally developing the high level of tension.

Controlling Motor Units

Motor units simplify the CNS control of the tension, or pulling force, produced by the fibers of skeletal muscle. An action potential transmitted down the axon of the motor unit activates all the fibers in the motor unit. The process of developing tension within a muscle is described in Chapter 15.

In this section, I first describe the basic relationship between an action potential activating a motor unit and the muscle tension produced. Then I describe two ways the CNS controls the amount of tension produced in a skeletal muscle through control of the multiple motor units.

Figure 14-6 shows the tension produced by a single motor unit (top) and the generation of an action potential in the motor unit (bottom) activating the muscle. A short vertical bar on the action potential portion of the figure indicates the generation of an action potential. The horizontal axis of both the force and action potential parts of the figure is time in milliseconds.

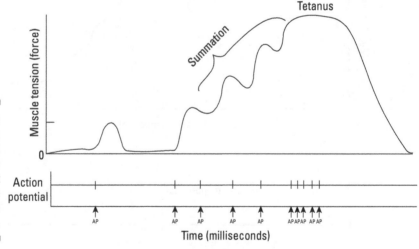

Figure 14-6:
The force produced by a single motor unit in response to action potentials.

Beginning at the left side, no action potential has been generated, and the muscle tension is 0. When an action potential is generated, the muscle responds with an increase in tension, called a *twitch,* as the fibers in the motor unit are activated. This single twitch response rises to a peak value and falls back to 0; this rise and fall occurs in less than 100 milliseconds, or one-tenth of a second.

After the tension falls back to 0, a new action potential is generated and the tension again increases. While the tension in response to the action potential is still present, another action potential is generated. The tension produced in response to this action potential increases the tension already present in response to the earlier action potential in a process called *summation,* or addition of the muscle tension. Summation continues when additional action potentials are generated before the tension begins to drop. Repetitive generation of action potentials leads to the muscle tension staying at a maximum level, a condition called *tetanus.* When no more action potentials are generated, the muscle tension declines back to 0.

To get more force from a muscle, the CNS uses an orderly system of activating motor units. Figure 14-7 shows the magnitude of muscle tension in the top part, and the generation of action potentials from five different motor units in the muscle in the bottom part. I use this figure to explain the two basic processes to increase muscle tension — motor unit recruitment and rate coding — in the following sections.

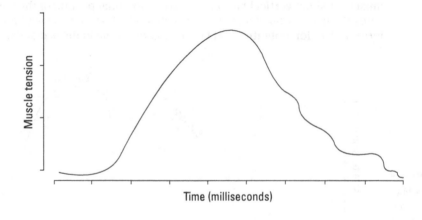

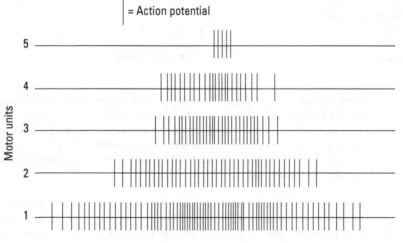

Figure 14-7: Motor unit recruitment and rate coding affect muscle tension.

Motor unit recruitment

One method of increasing the amount of muscle tension produced is to activate more motor units, a process called *motor unit recruitment*. Figure 14-7 shows that motor unit 1 is the first recruited, and the muscle tension begins to increase. As motor unit 1 continues to be active (action potentials continue to be generated), motor unit 2 is recruited. The muscle tension increases, as the tension contributed by both recruited motor units is summed. As motor units 3, 4, and 5 are recruited, the muscle tension increases even more.

Once recruited, a motor unit stays active as additional motor units are recruited. The amount of tension generated by a muscle increases as the tension contribution of each recruited motor unit is summed.

A decrease in muscle tension occurs in the opposite sequence of motor unit de-recruitment. Figure 14-7 shows that motor unit 5, the last motor unit recruited, is the first one to stop being active. Muscle tension decreases as individual motor units stop being active, and motor unit 1, the first motor unit recruited, is the last one to remain active.

Motor unit recruitment is used by the CNS to increase tension in all skeletal muscles. The recruitment process differs across various muscles. In some muscles, all the motor units are recruited by the time 50 percent of the maximum tension a muscle can generate is reached, and in other muscles motor unit recruitment continues until about 85 percent of maximum tension is produced.

The fact that all motor units can be recruited before maximum tension is produced leads to the second process used by the CNS to control the amount of tension produced.

Rate coding

Rate coding is the frequency at which action potentials are generated to activate a motor unit already recruited. Rate coding in skeletal muscle takes advantage of the summation of the tension produced by successive action potentials, as described in the previous section for a single motor unit.

In Figure 14-7, rate coding is evident in the changing frequency of action potential generation for the individual motor units 1 through 5. The space between the vertical bars indicates the time between generation of an action potential, and this is not constant for any of the motor units. As more motor units are recruited, the time decreases between successive action potentials generated in the already recruited motor units.

So, more tension gets contributed by each recruited motor unit even as more motor units are recruited. Because of rate coding, the tension contributed by each recruited motor unit can be increased.

Rate coding supplements motor unit recruitment to increase tension in all skeletal muscles. Each motor unit has a maximum rate of generating action potentials, ranging between 7 and 35 action potentials per second (yes, per *second* — the CNS and muscles are pretty quick).

The increase in tension reflects a combination of both motor unit recruitment and rate coding. Simplistically, a motor unit is recruited. As more tension is required, the recruited motor unit is activated more frequently. If more tension is required, the active motor unit generates action potentials more frequently (rate coding), and an additional motor unit is activated (motor unit recruitment). Still more tension needed? The CNS uses both rate coding and motor unit recruitment to generate the tension needed.

Chapter 15

Muscling Segments Around: Muscle Biomechanics

In This Chapter

▶ Looking at the characteristics of muscle

▶ Considering muscle structure

▶ Examining types of muscle activity

▶ Understanding muscle mechanics

▶ Zeroing in on the stretch-shorten cycle

---

*M*uscles are unique. They create tension, or pulling force, where they attach to the bones via tendons. Their activity is controlled by the central nervous system (see Chapter 14). In this chapter, I focus on skeletal muscle because these forces pull on and affect the motion of the individual segments of the body.

I begin the chapter by describing the general characteristics of skeletal muscle. Then I outline the structure of skeletal muscle responsible for the characteristics and explain how the structure creates both active and passive tension. I describe three types of muscle force production and relate them to the structure of muscle. The production of force by muscle is affected by the length of the muscle and by the rate of change of length of a muscle, and I explain these two relationships. Finally, I explain the stretch-shorten cycle, an effective way to increase the active and passive tension developed during a movement.

Characterizing Muscle

An individual skeletal muscle attaches to bone at each end by a tendon. There are 640 skeletal muscles in the body, and almost all cross at least one joint of the skeleton. At each joint, multiple muscles are grouped on each side of the joint. The muscles working as a group create a resultant force generating torque at the joint, as described in the link segment model presented in Chapter 8. For reference, major muscles of the skeletal system are shown in Figure 15-1. You can find more information on individual muscles in *Anatomy & Physiology For Dummies*, 2nd Edition, by Maggie Norris and Donna Rae Siegfried (Wiley).

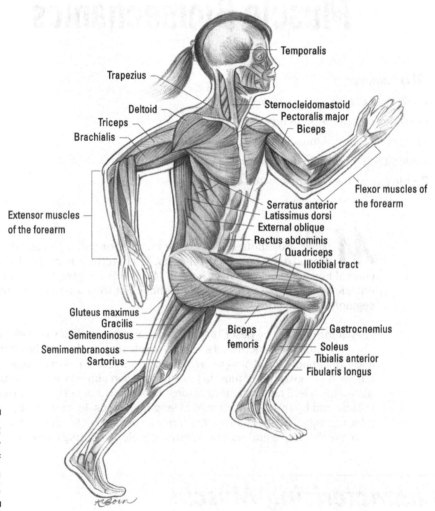

Figure 15-1:
Major
muscles of
the human
body.

A skeletal muscle consists of individual muscle fibers and connective tissue holding it together and attaching the muscle to the bones. All skeletal muscle exhibits the following four mechanical properties:

- ✔ **Extensibility:** A muscle can be lengthened or stretched beyond its initial length by an external force pulling on it. The external force can come from the opposing muscle group, gravity, or contact with another body. The connective tissue components allow the muscle to be stretched.

- ✔ **Elasticity:** A muscle wants to returns to its original length when the pulling force is removed because of the connective tissue components acting like a stretched spring. This is called *passive tension.*

- ✔ **Irritability:** Muscle responds to the stimulus of an electrical signal called the *action potential* delivered by the nervous system (see Chapter 14 for more on the action potential). When an action potential stimulates a muscle, it begins a process within the muscle fibers that causes the fourth property.

- ✔ **Contractility:** Muscle is a unique tissue because of its ability to develop active tension within the muscle fibers. In short, muscle converts the chemical energy taken in as food to a pulling force. I give an overview of the process of producing active tension later in the chapter.

Seeing How Skeletal Muscles Are Structured

The characteristics outlined in the preceding section are determined by the structure of skeletal muscle. Although the 640 individual skeletal muscles differ to some degree, they all have the same general structure of the typical muscle shown in Figure 15-2.

The structure of a "muscle" consists of two main parts: the muscle belly and a tendon at each end connecting the muscle belly to a bone. The muscle belly and tendon act as a mechanical unit to create and transmit tension to bone. To begin this section, I give a brief overview of the structure of the muscle-tendon (its *macrostructure*) and then describe the internal structure of muscle (its *microstructure*). I finish the section with a description of a three-component model of muscle. This model simplifies the anatomy yet still explains the four mechanical properties of muscle. The model has practical implications for explaining how a muscle and tendon apply force to affect movement.

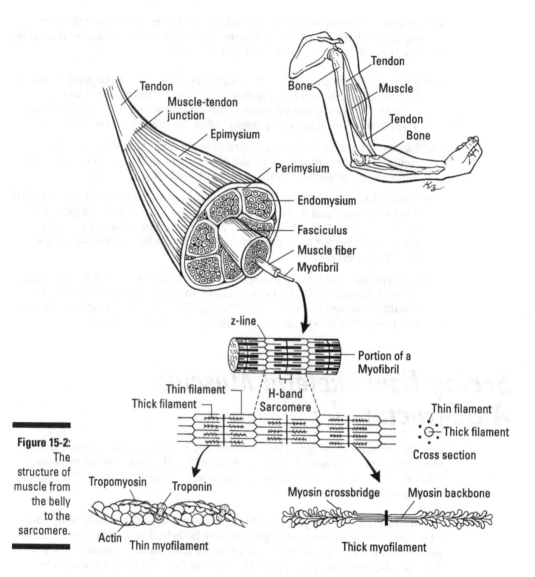

Figure 15-2:
The structure of muscle from the belly to the sarcomere.

The macrostructure of muscles

The overall structure of a muscle is shown in the cross section in Figure 15-2. I start the description at the innermost level and work towards the outer surface.

The muscle fiber is an individual muscle cell. Muscle force is produced within the muscle fiber. Muscle fibers are thin (much thinner than the thickness of a hair) and long (most go from tendon to tendon), like long, thin tubes.

The fibers, or tubes, are aligned parallel to each other. The fluid inside the muscle fiber is called the *sarcoplasm,* and the cell membrane surrounding the muscle fiber is called the *sarcolemma.*

A layer of connective tissue called the *endomysium* wraps around the fiber over the sarcolemma. The endomysium is a protective covering made of collagen and elastin fibers running the length of the fiber, and it also holds the fibers to each other. The endomysium joins into the tendon at the end of the muscle, or it may join to the endomysium of another fiber. There are blood vessels and nerve pathways within the endomysium. Blood brings nutrients into the fiber and removes waste. The nerves transmit signals to control and respond to muscle activity (see Chapter 14).

A *fascicle bundle* groups many parallel muscle fibers together. Each of the many fascicles in the muscle is wrapped in a layer of connective tissue called the *perimysium.* The perimysium holds adjacent fascicles to each other as it runs the length of the muscle belly to join into the tendon at each end.

The *epimysium* is an outer layer of connective tissue wrapping all the fascicles together to form the muscle belly. Like both the endomysium and the perimysium, the epimysium runs the length of the muscle belly to join into the tendons.

A tendon connects the muscle belly tight to bone at each end. The connective tissue of the endomysium, epimysium, and perimysium all join into the tendon, which anatomically is simply the three layers of connective tissue from the muscle belly extending beyond the ends of the muscle fibers to attach the muscle to bone. The sensory organ called the *Golgi tendon organ* (see Chapter 14) is intertwined with the connective tissue at the end of the muscle fibers where the separate tendon begins, called the *muscle-tendon junction.*

A tendon can be a rope-like structure like your biceps tendon, as in Figure 15-2, or it can be a broad band of connective tissue like your triceps tendon. Tendon contains mostly collagen and elastin along with some other fibers. The collagen fibers are arranged parallel to each other along the length of tendon to effectively transmit the pulling force of muscle to the bone. The pulling force is applied equally on the tendon at each end of the muscle.

The connective tissue layers within the muscle belly and the tendon can be stretched to store energy, much like stretching a spring. The storage and release of the energy within the connective tissue is called *passive tension,* and this tension affects the total force applied to the bones during movement, as I describe later in this chapter.

The microstructure of muscle fibers

Muscle fibers are the cells where muscle produces the pulling force. The development of pulling force within muscle is its unique characteristic. Muscles can only pull — they can't push. The pulling force is developed within the fiber in a structure called the *sarcomere* (refer to Figure 15-2). The force generated by the muscle is called *active tension,* and it's the sum of the force produced within individual sarcomeres.

Sarcomeres: The source of pulling force

An individual skeletal muscle fiber consists of a long line of sarcomeres connected end-to-end at what is called the *z-line,* as shown in Figure 15-2. Within each sarcomere is a three-dimensional arrangement of thick and thin protein filaments. The thin filaments are attached to the z-line at each end and project into the central region of the sarcomere. The thick filaments are fixed in the central region of the sarcomere by a structure called the *H-band.* Each thick filament is surrounded by six thin filaments, as shown in the cross section in Figure 15-2. The overlapping of the thick and thin filaments give skeletal muscle a striated appearance of alternating light and dark bands when viewed under a microscope.

The thin filament is actually a two-stranded filament made of the protein actin. The two strands are braided together. Along each of the strands are binding sites, covered by a protein called *troponin.* When an *action potential* (an electrical signal coming from the central nervous system; see Chapter 14) travels along the sarcolemma, it causes an electro-chemical reaction of the membrane, and calcium is released into the sarcoplasm. The calcium binds with the troponin and pulls it out of the way to uncover the binding site. The thin filament is now ready to interact with the thick filament.

The thick filament is made of the protein myosin. The filament appears thick because of projections called the *crossbridge* aligned in opposite directions along each end of the backbone of the filament (see Figure 15-2). When the binding sites on the actin are uncovered, the myosin crossbridges bend out from the backbone of the myosin, and a chemical bond is created between the crossbridge and the binding site on the thin filament. The crossbridge then swivels, pulling the thin filament towards the center of the sarcomere in an action called the *power stroke.* Because of the arrangement of the crossbridges on the thick filament, the pull of the power stroke is always toward the center of the sarcomere; muscles can only pull. The bond between the crossbridge and the binding site is broken by another chemical reaction. The crossbridge moves back toward the backbone of the myosin, and the cycle of power strokes continues as long as the muscle is being stimulated by an action potential.

The process of developing tension within the sarcomere described here is called the *sliding filament model of muscle force production* and results in active tension.

When a muscle fiber is activated, thousands of the power strokes are created, each pulling on the thin filament. The power strokes don't all occur at the same time, so the thin filaments are pulled toward the center by the thick filament throughout the period of muscle activation. The net effect of all the power strokes is to try to pull the z-lines of the sarcomere toward each other. Because all fibers in a single motor unit and multiple motor units are recruited at the same time (I explain the motor unit and motor unit recruitment in Chapter 14), the force actively generated by the muscle and transmitted through the tendon to bone is the summed effect of many thousands of individual power strokes.

Simplifying muscle: A three-component model of muscle

Because the connective tissue layers and the tendon work with the muscle fiber to develop, store, and transmit force to bones, it's common to refer to the muscle, connective tissue layers, and tendon as one unit called "the muscle." The complex anatomy of the muscle can be simplified into a three-component model that explains the muscle properties of extensibility, elasticity, and contractility under the control of a stimulus from the central nervous system (irritability). Figure 15-3 shows one version of the three-component model of muscle. Here are the components:

- ✔ **Contractile component:** The contractile component represents the unique force-generating ability of muscle. It's drawn as a simplified sarcomere, showing a single thick filament and four thin filaments. Interaction of the thick and thin filaments is the source of active force production by the muscle.

- ✔ **Series elastic component:** The series elastic component represents the tendon. It's drawn as a spring in line with, or in series with, the contractile component of the simplified muscle.

- ✔ **Parallel elastic component:** The parallel elastic component represents all the connective tissues surrounding the various levels or bundles of the anatomic muscle. It's drawn as a spring parallel to, or beside, the contractile component.

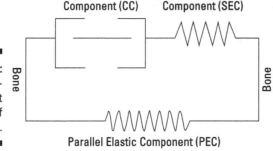

Figure 15-3: A three-component model of muscle.

The three-component model still explains the elasticity, extensibility, and force-generating capability of the muscle. The spring-like parallel and series elastic components account for how the muscle can be stretched passively (extensibility) and then return to its original shape (elasticity). The contractile component represents the unique ability of muscle to develop tension (contractility).

Comparing Types of Muscle Activity

Muscle acts as a torque generator at the joints (see more on torque and muscle force as a torque generator in Chapter 8); the muscle pulls on the segments affecting rotation at the joint the muscle crosses. Consider the simple link segment model (see Chapter 8) of the upper arm and forearm-hand segments shown in Figure 15-4. A single muscle crosses the front of the elbow joint. The muscle has fixed attachments to the upper arm and the forearm — the distance between the attachments is the length of the muscle. The force from the muscle creates a muscle torque in the counterclockwise direction, tending to flex the forearm. A handheld weight (HHW) creates a loading torque in the clockwise direction, tending to extend the forearm (note that segment weight is ignored in this example). I use this simple model to explain isometric, concentric, and eccentric muscle activity.

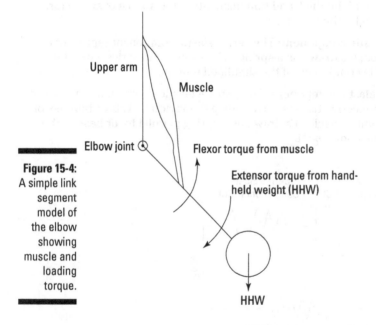

Figure 15-4:
A simple link segment model of the elbow showing muscle and loading torque.

Upper arm

Muscle

Elbow joint

Flexor torque from muscle

Extensor torque from hand-held weight (HHW)

HHW

I use the word *activity* to describe the development of tension within a muscle because a word commonly used, *contraction,* means "to shorten." There are three types of muscle activity, but only one involves shortening of the muscle while it develops tension.

Isometric activity

A muscle is active isometrically when it develops tension but no movement occurs at the joint it crosses. No movement occurs because the muscle torque generated by the muscle is offset by the loading torque in the opposite direction. The distance between the attachment points of the muscle stays the same length.

In Figure 15-5, the joint motion is 0 at points A, B, and C. The muscle is active at A, B, and C, producing the force to generate a flexor torque equal in magnitude to the extensor torque generated by the HHW. Because the joint motion is 0, the muscle is active isometrically. The muscle is active, but no movement is occurring at the joint it crosses.

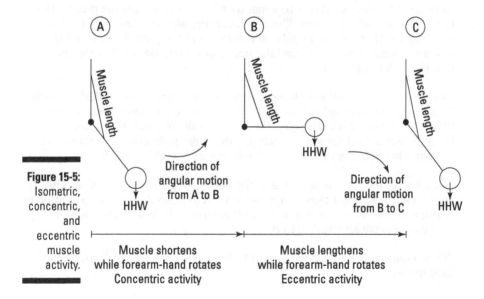

Figure 15-5: Isometric, concentric, and eccentric muscle activity.

During isometric activity, the loading torque can be generated by an external force applied to the segments meeting at the joint (like the HHW above) or by the muscle group on the opposite side of the joint. When a muscle is active isometrically, it prevents joint movement.

Isometric activity is sometimes defined as follows: "The muscle develops tension, but it does not change length." Actually, the muscle fibers get shorter during isometric activity as they develop active tension, but the shortening of the fibers is offset by a lengthening of the elastic components within the muscle, mainly the tendon. The muscle-tendon unit as a whole stays the same length, not the individual muscle fibers.

Concentric activity

A muscle is active concentrically when it develops tension and gets shorter in length. The attachment points of the muscle get closer together. The torque generated by the active muscle as it shortens is in the same direction as the angular motion.

Consider the motion of the link segment model from A to B in Figure 15-5. At both A and B, the segments are at rest and the flexor muscle torque is equal in magnitude to the extensor loading torque.

To start the flexing motion and move the forearm from A to B, the muscle force must be increased at A to generate a flexor muscle torque greater than the extensor loading torque. The net flexor torque (for more on net torque, see Chapter 8) causes an angular acceleration of the joint flexion, and the forearm begins to rotate toward the upper arm (angular acceleration is described in Chapter 9).

To stop the flexing motion to bring the forearm segment to rest at B, the muscle force must be decreased so the flexor muscle torque generated becomes less than the extensor loading torque. The net extensor torque, acting opposite to the direction of the joint flexion, causes a decrease in the angular rotation of the forearm toward the upper arm, and the forearm comes to rest at B.

From A to B, the muscle is active as the fibers are producing tension. The attachment points get closer together from A to B, so the muscle gets shorter. Because the muscle is active while getting shorter in length, the muscle is active concentrically from A to B.

When a muscle is active concentrically, the muscle is causing the observed joint motion.

Eccentric activity

A muscle is active eccentrically when it develops tension and gets longer. The attachment points of the muscle get farther apart. The torque generated by the active muscle as it lengthens is opposite to the direction of the angular motion.

Consider the motion of the link segment model from B to C in Figure 15-5. At both B and C, the segments are at rest and the flexor muscle torque is equal in magnitude to the extensor loading torque.

To start the extending motion that moves the forearm from B to C, the muscle force must be decreased at B so the flexor muscle torque generated is less than the extensor loading torque. The net extensor torque (see Chapter 8) causes an angular acceleration of joint extension, and the forearm begins to rotate away from the upper arm.

To stop the extending motion so the forearm segment is at rest at C, the muscle force must be increased so the flexor muscle torque generated becomes greater than the extensor loading torque. The net flexor torque, acting opposite to the direction of the joint extension, causes a decrease in the angular rotation of the forearm away from the upper arm, and the forearm comes to rest at B.

From B to C, the muscle is active as the fibers are producing tension. The attachment points get farther apart from B to C, so the muscle gets longer. Because the muscle is active while getting longer, the muscle is active eccentrically from B to C.

When a muscle is active eccentrically, the muscle is controlling movement; an external force creates the torque causing the movement, and the resisting torque from the muscle force controls or slows down the moving segment. In the case earlier, the torque from the HHW is causing the extension of the forearm at the elbow, and the muscle torque acts to control the initial rate of joint extension and then stop the extension at point C.

When a muscle is active and the joint doesn't move, the muscle activity is isometric. Isometric muscle activity creates a muscle torque to prevent joint motion.

When a muscle is active and the joint moves in the same direction of the muscle torque, the muscle activity is concentric. Concentric muscle activity creates a muscle torque that causes joint motion.

When a muscle is active and the joint moves in the opposite direction of the muscle torque, the muscle activity is eccentric. Eccentric muscle activity creates a muscle torque that controls and stops joint motion.

When learning about individual muscles, the role of the muscle is often described as *flexor* or *extensor, adductor* or *abductor,* and *internal rotator* or *external rotator* (see the description of individual muscles in *Anatomy & Physiology For Dummies*). For example, the muscles on the front of the forearm, simplified in the preceding example, are described as "elbow flexors." It's better to think of the torque created by an individual muscle rather than the action caused by a muscle. The isometric, concentric, and eccentric activity of a muscle is equally important — preventing, causing, and controlling/stopping motion are all part of the role of an individual muscle.

The amount of force required of a muscle can change during a joint movement as the torque required from the muscle changes. During flexion of the elbow joint described earlier, the magnitude of the flexor muscle torque changed from greater than the loading torque to start the joint flexion to less than the loading torque to stop the joint flexion. The amount of activation in the muscle is affected by how many and how often the individual motor units are recruited, as described in Chapter 14.

In the next section, I describe additional factors affecting the force produced by a muscle when it's activated by the central nervous system.

Producing Muscle Force

A muscle fiber responds to an action potential (the stimulus) by developing tension, a pulling force, and transmits the force through tendons to the bones. The amount of tension required from a muscle depends on the magnitude of the torque needed to cause, stop, or control motion at the joint, as explained earlier in this chapter and in Chapter 8. The amount of tension a muscle can produce to generate the torque depends on several factors. I describe two of these factors in this section.

Relating muscle length and tension

The relationship between the length of the muscle and the amount of tension produced is called, not surprisingly, the *length–tension* (L-T) *relationship*. The total tension, or pulling force, comes from the contractile, series, and elastic components (the three-component model of muscle explained earlier in this chapter and shown in Figure 15-3). Figure 15-6 shows the L-T relationship for the individual components and for the total muscle, with length on the horizontal axis and tension (force) on the vertical axis. I explain the relationships in this section.

Relating length and tension in the contractile component

The unique characteristic of muscle is its ability to produce force from the power strokes produced between the thick and thin filaments in the sarcomere. In the three-component model of muscle, the power strokes are produced in the contractile component.

The pulling force produced by the contractile component is called *active tension*.

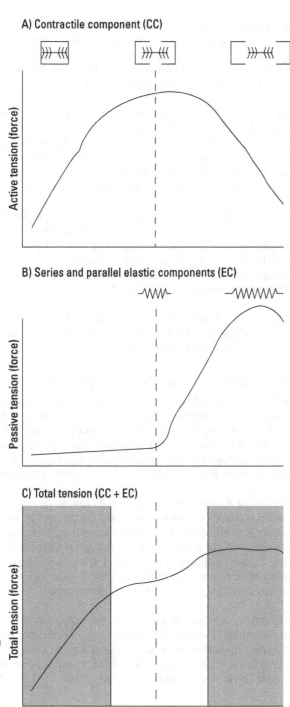

A) Contractile component (CC)

Active tension (force)

B) Series and parallel elastic components (EC)

Passive tension (force)

C) Total tension (CC + EC)

Total tension (force)

Shortened

Resting length

Lengthened

Figure 15-6: The length–tension relationship for muscle.

More power strokes produce more active tension. For power strokes to occur, the thick and thin filaments must overlap to allow the crossbridges to link at the binding sites. The optimal overlap occurs when the sarcomere is at what is called *resting length*. And resting length is where maximum active tension is produced (see the top part of Figure 15-6).

If the muscle, and the contractile component, is lengthened, there is less overlap between the thick and thin filaments. Active tension decreases as the contractile component is lengthened, as shown in Figure 15-6. When the filaments have no overlap, there can be no active tension produced by the contractile component when the muscle is stimulated because no cross-bridges can link to binding sites and no power strokes can be developed.

If the muscle, and the contractile component, is shortened, active tension also decreases. This happens because the way a crossbridge projects from the thick filament allows it to link only to a binding site on the thin filament at the same end of the contractile component. As shown in Figure 15-6, as the contractile component gets shortened from the resting length, the thin filaments begin to overlap and fewer binding sites are available.

Lengthening or shortening of the contractile component decreases the number of potential binding sites for the production of power strokes, decreasing the amount of active tension produced.

Relating length and tension in the elastic components

Muscle is extensible and elastic because of the connective tissues around the layers and in the tendon, represented by the parallel elastic component and the series elastic component of the three-component model of muscle. Although they're separate structures, the parallel elastic component and series elastic component respond the same way to any change in muscle length, so I'll combine them into one unit called the elastic component.

When a force pulls on the elastic component, it gets longer (stretches) like the spring on a screen door (that's why it's represented in the model as a spring). Strain, or elastic, energy is stored in the lengthened elastic component (strain energy from the deformation of a material is explained in Chapter 7). When the force stretching the elastic component is released, the stored elastic energy pulls the component back to its original shape. The tension, or pulling force, produced in the elastic component is called *passive tension*.

The parallel elastic component and series elastic component actually contribute to muscle tension differently. The parallel elastic component only contributes tension during eccentric muscle activity, when the distance between the muscle attachment sites gets farther apart. The series elastic component contributes during isometric, concentric, and eccentric muscle activity.

The L-T relationship for the elastic component is shown as the middle part of Figure 15-6. At the initial length, the elastic component is slack and has no stored energy to convert to tension. Similarly, if the elastic component is shortened, it just gets slacker and no energy is stored. As the elastic component is stretched during lengthening, the stored energy increases. More stretch = more stored energy = more tension produced when the stretched elastic component is released.

Lengthening the elastic component increases the amount of passive tension produced.

Relating length and tension in the total muscle

The complex anatomy of a whole muscle is simplified in the three-component model. In a real muscle, neither the contractile component nor the elastic component acts alone; both contribute to the tension produced by the muscle. The total tension produced is the sum of the tension from the contractile component and the elastic component. The L-T relationship for the total muscle is shown in the bottom of Figure 15-6. This L-T curve is the sum of the contractile component and elastic component curves, and includes both active and passive tension.

Because it's attached to bones at each end, the length of a muscle crossing a joint stays in the unshaded region of the L-T relationship. The total tension produced is affected by the length of the muscle, and the total tension is higher when tension comes from both the contractile component and the elastic component. I explain how the L-T relationship applies to effective joint motion later in this section.

More total muscle force is produced if the muscle is elongated, or stretched, because both active and passive tension contribute to the total force.

Relating muscle velocity and tension

The power stroke comes from links created between the thick and thin filaments in the sarcomere, as I describe earlier. Power strokes are created in isometric, concentric, and eccentric muscle activity. The velocity of the change of muscle length — how fast it shortens or how fast it lengthens — affects the amount of muscle force (tension) produced. The relationship is described by the *force–velocity* (F-V) *relationship*. Figure 15-7 shows the F-V relationship for eccentric, concentric, and isometric activity, with muscle velocity on the horizontal axis and muscle force (tension) on the vertical axis.

The amount of muscle force is highest in eccentric activity, lowest in concentric activity, and in the middle during isometric activity.

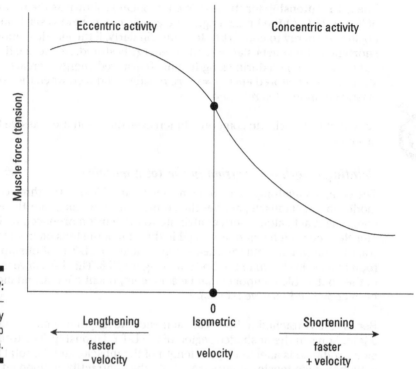

Isometric activity: Relating muscle force to zero velocity

In isometric activity, there's no increase or decrease between the attachment points of the muscle; the muscle velocity is zero. The force produced isometrically is shown in the middle of Figure 15-7.

During isometric activity, the thick and thin filaments of the sarcomere don't move very much during the power stroke. Because the crossbridge pulls on a relatively motionless thin filament, the link is maintained for a relatively long time and a high level of pulling force is produced before the link is broken. Also, many links can be created at the same time with each contributing to the active tension produced. Basically, many high-force links are produced during isometric muscle activation.

Concentric activity: Relating muscle force to shortening

In concentric activity, the muscle shortens while it develops tension. The F-V curve shows the force is less than in isometric activity, and actually decreases with faster shortening velocity, as shown in Figure 15-7.

During concentric activity, the thin filaments are sliding over the thick filaments toward the center of the sarcomere. Because the thin filament is moving, some binding sites slide past available crossbridges of the thick filament before a link can be created. Fewer links get created during muscle shortening than during isometric activity. When a link is created, the crossbridge pulls on the thin filament in the direction it's already moving. The pulling force of the crossbridge at the binding site moving in the direction of the pull is not as high as can be produced when the binding site is not moving, and the link cannot be maintained for as long a time as in isometric activity. With a faster shortening velocity, even fewer and less forceful power strokes are created for a shorter period of time. Basically, fewer and lower force exerting links are produced during concentric muscle activation than in isometric muscle activity.

Eccentric activity: Relating muscle force to lengthening

In eccentric activity, the muscle lengthens while it develops tension. The F-V curve shows the force is greater than in isometric activity, and the force actually increases with faster velocity of lengthening.

During eccentric activity, the thin filaments are sliding over the thick filaments but away from the center of the sarcomere. As in concentric activity, some binding sites are missed. However, when a link is created, the crossbridge pulls on the thin filament in the direction opposite to the direction the filament is sliding. The pulling force of the crossbridge on the binding site moving in the opposite direction is even higher than when the binding site is not moving as in isometric activity. With a faster lengthening velocity, fewer but even more forceful power strokes are created. Basically, more and higher force-creating power strokes are produced during eccentric muscle activation than in isometric muscle activity.

Stretching before Shortening: The Key to Optimal Muscle Force

Muscle force serves as a torque generator at joints. Muscle produces more force when both active and passive tensions contribute. An effective way to ensure that both active and passive tension contribute is to use the stretch-shorten cycle.

The stretch-shorten cycle involves quickly elongating a muscle (the stretch) before it shortens to create torque at the joint it crosses — hence, the name *stretch-shorten cycle*. I explain the joint motions that characterize the SSC using Figure 15-8, representing a link segment model of a leg during a vertical jump.

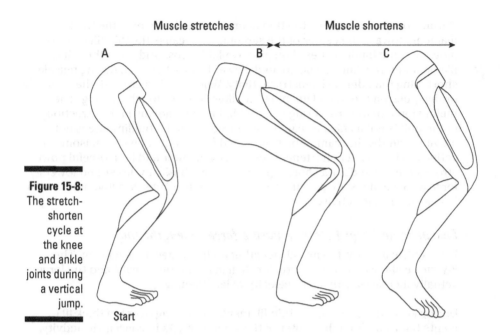

Muscle stretches   Muscle shortens

A   B   C

Figure 15-8:
The stretch-
shorten
cycle at
the knee
and ankle
joints during
a vertical
jump.

Start

In position A of Figure 15-8, the performer is at rest. The knee and ankle joint angles are kept in position by isometric activity in the two muscles shown. To push down on the ground to perform a vertical jump, the jumper uses concentric activity in the muscle on the front of the knee to *extend* (straighten) the knee and concentric activity in the muscle on the back of the ankle to *plantarflex* (straighten) the ankle joint. These actions create the push down on the ground. If the performer starts the upward motion, or push-off, from position A, the muscle force comes only from active tension produced during the concentric activity.

A large torque is desired from the muscles at the hip and at the knee joints during push-off to accelerate rotation of the lower leg and foot to create a large upward force from the ground to push the performer upward during the jump.

To use the stretch-shorten cycle, the performer first allows the knee to bend (flex) and the ankle to bend (dorsiflex). During these joint motions, the muscles on the front of the thigh and on the back of the ankle are lengthened and become more active eccentrically to stop the joint motion at position B. This represents the stretch phase of the cycle. Between B and C, the knee is straightened (extended) and the ankle is plantarflexed (straightened) by the torque developed by the concentric muscle activity. This is the shorten phase of the cycle. The muscle force during the shorten phase of the stretch-shorten cycle includes both active and passive tension.

The stretch before shorten phasing of the stretch-shorten cycle produces more muscle force and more torque during the shortening phase because the lengthening phase serves to:

- ✔ **Increase the time the muscles develop force:** Because the contractile component of the muscle is active eccentrically during the stretch phase of the stretch-shorten cycle to stop the motion of the joint, the active force contribution is already high when the muscle shortening begins.

- ✔ **Store elastic energy in the muscle as the elastic component is elongated:** The stored energy contributes passive tension to the total force produced by the muscle during the shortening phase of the stretch-shorten cycle.

- ✔ **Invoke the stretch reflex as the muscle spindles are deformed during the stretch phase (Chapter 14):** The stretch reflex enhances the activation of the muscle to create more active tension by the contractile component.

There are two key requirements to enhancing muscle force when using the stretch-shorten cycle:

- ✔ **The stretch phase should be rapid to require more eccentric activity to stop the joint motion and to invoke the stretch reflex.** The rate at which the muscles are stretched is more important than by how much the muscles are elongated.

- ✔ **There should be no pause between the stretch and shorten phases so that the stored energy and the high level of active tension are both utilized during the shortening of the muscle.** Energy cannot be stored for a long time in the elastic component — if the pause is too long, the muscle length is adjusted by the nervous system (the muscle spindle and Golgi tendon organ responses, as explained in Chapter 14). Muscle shortening would produce only active tension.

The stretch-shorten cycle is very common in skilled human movements. It's the basis of including a windup during skills like throwing and kicking, the backswing in golf, and a countermovement in jumping, as described earlier. It's used in walking and running, when the joints of the leg are slightly flexed following initial foot contact with the ground and then extended during the push-off before the foot comes off the ground.

Regardless of what it's called, the stretch-shorten cycle increases the total tension exerted actively and passively by the muscles during the shorten phase of the skill.

Part V
Applying Biomechanics

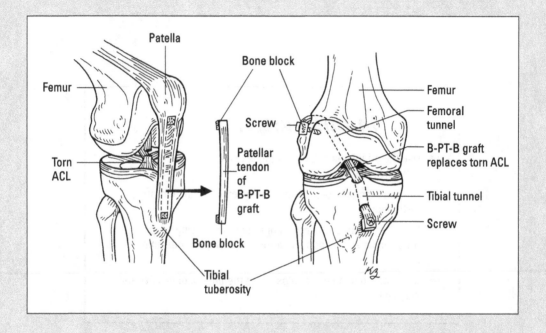

Find out how to turn around in outer space in a free article at http://www.dummies.com/extras/biomechanics.

In this part...

- Use the principles of biomechanics to identify and correct movement performance.

- Measure motion, force, and muscle activity with laboratory instruments.

- Use biomechanics to know more about exercising in space, rebuilding a knee, choosing a running style, and designing a helmet.

- See how biomechanics helps get to the bottom of murder and mayhem.

Chapter 16

Eyeballing Performance: Qualitative Analysis

A qualitative analysis involves systematically observing and evaluating the quality of a movement. Instead of measuring the performance using instruments and reporting numbers, a qualitative analysis describes the quality of the performance using words such as *more* or *less, longer* or *shorter, faster* or *slower, higher* or *lower, larger* or *smaller, harder* or *softer,* and *forward* or *backward.*

In most cases, a qualitative analysis involves judging the level of performance and providing feedback to the performer. The purpose of a qualitative analysis is to help the individual reach a satisfying level of performance.

A qualitative analysis should focus on the factors that determine the performance level. These factors relate to the production and application of force. In this chapter, I provide a general overview of conducting a qualitative analysis and give an example using a relatively common movement, throwing.

Before I get started, just one note: I call anyone performing a movement "the performer" and anyone evaluating and giving feedback regarding the quality of a performer's movement a "movement analyst."

Serving as a Movement Analyst

Any eyeball evaluation of performance represents a qualitative analysis. The performance may be viewed in real-time, or it may be a recorded performance that is analyzed later. Regardless, the evaluation is based solely on what the analyst can observe.

The qualitative analysis can be informal, like the expressed awe of a spectator watching the gold medal–winning performance of an Olympic athlete. The awe expressed is a qualitative evaluation of movement, even though the evaluation will likely never get back to the performer.

The qualitative analysis can be more formal, such as when an instructor or parent works with a child learning to throw, when a clinician works to rehabilitate the gait of an injured worker or stroke victim, or when a coach works with an athlete to improve the high jump. In this type of qualitative analysis, there's typically an expectation that the movement analyst will evaluate the quality of performance and provide feedback to the performer on how to "move better." I focus on this more formal qualitative analysis in this chapter.

A movement analyst has three responsibilities when conducting a qualitative analysis:

- ✔ **Evaluate:** The analyst must watch the performance and judge whether the goal, or objective, of the movement is attained. Basically, this means answering the question "Is the performer doing what he wants to do or is supposed to do?" If the goal is attained, then new goals are set. If the goal isn't attained, then the insight of the movement analyst is needed to help improve performance. One tricky part in determining whether the goal is reached comes in actually specifying the goal of the performance, as I explain in the next section.

- ✔ **Troubleshoot:** The analyst must pinpoint the reason the goal is not attained. This part of a qualitative analysis is the most important. It involves the accurate identification of the *specific cause* of not attaining the performance goal.

- ✔ **Intervene:** The analyst must correct the cause of the performer not attaining his performance goal.

Unfortunately, many movement analysts evaluate a performance and intervene without actually troubleshooting the precise cause of a less-than-satisfying performance. Instead of correcting the specific cause of a problem, the analyst uses well-intended phrases such as "Good try," "Jump higher," "Throw farther," or "Run faster." But these phrases don't provide a lot of assistance for the performer, because they don't get to the specific cause of the lower jump, shorter throw, or slower run (although many people of all ages appreciate having effort acknowledged with a "good try").

The appropriate intervention to improve performance depends on what is typically the most overlooked step: troubleshooting the cause of the performance as exhibited. In the following sections, I provide an overview of conducting a qualitative analysis of movement, including troubleshooting based on the biomechanics covered in Parts II, III, and IV of this book.

Evaluating the Performance

Evaluating the performance consists of three steps: (1) Identifying the goal of the movement, (2) Specifying the mechanical objective, and (3) Determining whether the goal has been met. I explain each of these steps in this section.

Identifying the goal of the movement

Knowing the goal of the movement is a critical step in a qualitative analysis. With a specified goal in mind, the performance can be evaluated to determine whether the goal is reached. If the performer fails to attain the goal, the cause of the failure must be identified and corrected.

The goal of a movement is its intended outcome. A clear statement of the specific goal of a movement is needed in order to evaluate the level of performance. This seems like an easy task, but it can get complicated because there are two broad categories of movement goals: the *product* (the outcome of a movement) and the *process* (how the movement is performed). Evaluating whether the goal is reached is difficult without a clear specification of which goal is desired.

Producing the outcome

The product is what the performer accomplishes by the movement. Many movements have an intended outcome, which I call the *strategic goal* (sometimes referred to as the *criterion* or the *objective*). Sometimes, performance is evaluated based solely on attaining the strategic goal. It's assumed that good technique is used by the performer, or, less frequently, how the goal is accomplished doesn't matter as long as the goal is attained.

When shooting a basketball, the strategic goal is usually to get the ball through the hoop. (This goal can be adapted for any throwing, kicking, or hitting task — there's a desirable target the performer aims for.)

In walking, the strategic goal is to get from point A to point B. (This can be adapted for jogging, running, skating, skipping, or any form of locomotion — there is a desired change in location. There may be a time aspect, too, like running fast in a race or walking slow to enjoy a fine day outdoors.)

The product of a performance can be compared to the strategic goal to determine whether the movement was successful.

Processing the motion

The process is how the movement is performed. Other common names for how a movement is performed include *style, technique,* and *form.*

I use the term *process* because movement is created by a process of manipulating force. The link-segment model of the body (see Chapter 8) represents the individual segments of the body as rigid links, or sticks, meeting at joints where the segments rotate. Muscles cross the joints, and the pulling force of muscle creates a muscle torque on the segments.

The torque from the muscles crossing the joints accelerates the segments, changing the segmental angular velocity at the joints. Muscle force from the performer controls the rotation of the segments:

- The performer uses the motion of the rotating segments to produce an external force, usually from the ground, to control the change in the motion of her body when walking, running, jumping, or landing.

- The performer uses the motion of the rotating segments to apply an external force to an object in movements like throwing (for example, a ball, a javelin, or a rock), weightlifting (for example, the barbell, dumbbell, or machine handle), or striking an object with an implement (where the bat, racket, or stick becomes an extension of the performer's body segments).

The rotating motion of segments created by muscle torque throughout the movement produces the external force and creates the product. In general, faster angular motion of more segments produces a larger external force. Using isolated segment or body positions at specific instances as indicators of "good" performance overlooks how the segments got into the positions — the process of performing the movement. The positions are simply instants during the correct process, and they occur because of the correct angular motions that moved segments through the positions.

The process (kinetics) of the movement determines the product (kinematics) of the movement. An unbalanced force causes a change in motion.

Specifying the mechanical objective

Recognizing what the performer controls while performing a skill is a big step to improved movement analysis. Any feedback during the intervention should be directed to factors the performer controls. This seems like a pretty obvious point, but it's ignored a lot. Intervention must be focused on the cause (the process) and not the effect (the product).

In movement, force is the cause and motion is the effect, as emphasized by Newton's laws (see Parts II and III of this book). I like to use the impulse–momentum relationship (derived from Newton's second law, covered in Chapters 3 and 6) as the basis of defining the mechanical objective, although the work–energy relationship (explained in Chapter 7) could also be used.

The strategic goal of a movement requires getting a body (the entire body in locomotion; a ball in a striking or throwing activity) to move at a desired speed in a desired direction. This is the momentum of a body (momentum = mV). The strategic goal is what the performer wants to accomplish with the body, and the product is what we "see" happen to the body, what the performer actually accomplishes. The body evaluated has a mass, and the strategic goal is reached if the performer uses angular motion of the segments to produce an external force to get the body to move at a particular speed in a particular direction. Looking at movement this way, the strategic goal of a performer is to apply a force to change the momentum of a body to a desired or required level.

The process is what the performer controls to cause the change in momentum. As explained earlier, the process involves applying force to the body to produce the correct momentum; the force applied to a body for a period of time is the impulse (impulse = Ft). The performer uses muscle force to produce angular motion of the segments and uses the angular motion of the segments to generate an external force to get the body moving with the momentum needed for success. From a mechanical perspective, the angular motion of the segments is used to apply an impulse to the body — a force is applied for a period of time.

Impulse causes a change in momentum. If the momentum is incorrect, if the body doesn't move as desired and the strategic goal is not attained, the problem stems from the impulse generated.

The mechanical objective of any skill is for the performer to optimize the impulse applied to the body to be moved. *Optimize* means to tailor the impulse produced to the specific situation — sometimes you need a lot of impulse (for example, to jump high), sometimes you need a little impulse (for example, to jump low), and sometimes you need a medium-size impulse (for example, to jump to a height in between). The applied impulse must match the amount of momentum required to attain the strategic goal.

The equation for impulse is *Ft*, or force applied for time. This relatively simple equation provides a mechanically sound focus for identifying the possible sources of a problem with performance:

- ✔ **The amount of force applied:** The magnitude of the external force can be too much or too little, causing a dissatisfying performance.

- ✔ **The direction of the force applied:** The external force can be applied in the wrong direction from the intended momentum, causing a dissatisfying performance.

- ✔ **The duration, or time, of force application:** The external force can be applied for too short or too long a time, causing a dissatisfying performance.

With a two-term equation explaining the cause of motion, life is good for a movement analyst. But the challenge comes in using *Ft* to improve performance.

The process is the application of impulse, and the product is the resulting momentum of the body. The size, direction, and duration of force applied by the rotating segments are the cause of the observed motion of the body; the process of applying impulse causes the product of the observed momentum.

Successful performance depends on the application of the optimal impulse in the correct direction. Any combination of errors in size, direction, or duration of force application creates incorrect impulse and causes incorrect momentum, and a dissatisfying performance.

Determining whether the goal has been reached

When you know what the goal is, you're ready to evaluate whether the goal — whether it's the product or the process — has been met.

Sometimes, even when the performer's strategic goal is attained, the movement analyst judges that "something isn't right" — the style, technique, or form doesn't match with the analyst's "mental image" of how the skill is supposed to look. The mental image is often based on body and segment positions exhibited at specific instants during performance — typically the start, the end, or isolated instants between the start and end positions.

In throwing, for example, critical positions of the "mental image" include "a leading elbow" position of the throwing arm (shown in Figure 16-1) and evidence of a "follow-through" after ball release (shown in Figure 16-2). If these desired isolated positions aren't seen, the analyst may feel that the performer's technique needs some work.

Figure 16-1:
The leading elbow position of the arm when throwing.

The problem is that the mental image is an instant during the performance and doesn't consider how the body or segments got into those positions. Asking the performer to exhibit the positions without focusing on how the segments moved into and then through the desired positions won't improve performance. The positions are symptoms of a good performance, not the cause of a good performance. Similarly, failure to attain the positions is a symptom of poor performance, not the cause of poor performance. I come back to this later in the chapter.

When both the analyst and the performer agree that the performance should be improved, the analyst must troubleshoot the performance to identify the cause of the less-than-satisfying performance. And that's the subject of the next section.

Figure 16-2:
The follow-through position during a throwing motion.

Deciding what's "good enough"

An entire philosophical dilemma is associated with goal attainment that goes beyond biomechanics. Why does the person have to "move better" anyway? Isn't moving on its own, regardless of performance level, a justifiable goal because of the health benefits associated with physical activity? The benefits of physical activity come despite the level of performance. If someone enjoys an activity despite a short-coming in success, most will continue to do it and get the benefits from being physically active. Well-intended but unwelcomed feedback may cause someone to quit an activity and, in doing so, become less physically active. The advice of a movement analyst is warranted only if a performer welcomes the evaluation.

The performer's level of satisfaction should be the determinant of whether any intervention is needed, even if the movement analyst may have concerns about how the performer attained the strategic goal. If the performer is satisfied with the movement, the performer can be encouraged to continue. No intervention is required.

For example, if an elderly stroke victim attends rehabilitation sessions to regain the ability to move independently, this performer may be satisfied with being able to get around with the assistance of a cane or a walker. The performer may be ready to get on with life accepting this assisted, but independent, form of locomotion. The performer may feel he has better things to spend time on than enduring the additional grueling rehabilitation sessions necessary to get around without the cane or walker.

Troubleshooting the Performance

Troubleshooting is all about identifying the *source* of the dissatisfying performance. The level of performance may be less than satisfying because of constraints or because of technique errors. Improvements in performance and satisfaction with the performance occur when the constraints or technique errors are remedied. I begin this section by explaining constraints; then I move on to technique errors.

Constraints on performance

A *constraint* is something that restricts, or constrains, the level of performance. Constraints can come from three sources:

- ✔ **The task:** Task constraints are specific to the movement performed. The primary task constraint is the strategic goal of the task, but task constraints also include rules specifying how, and with what, a movement must be performed. For example, in baseball or softball the pitcher must throw from the rubber of the mound, in basketball the ball must go through the hoop, and in walking the performer is generally expected to walk without assistance.

- ✔ **The environment:** Environmental constraints are external to the performer and, as the name implies, come from the location or environment in which the movement takes place. Environmental constraints include obvious factors like surfaces, lighting, temperature, and acoustics, but also the presence or absence of other players and observers, or even strategic situations within a game. Critical but often overlooked environmental constraints include the selected, not regulated, dimensions (height, width, length) of the performance space and the equipment used (dimensions and mass of implements and any targets).

- ✔ **The individual:** Individual constraints are internal to the performer, reflecting the attributes of the performer's neuromusculoskeletal system. Individual constraints are either structural or functional:

  - **Structural constraints** are determined by the performer's dimensions, including body height, segment lengths, and body mass. A physical condition like paralysis, limb loss, or joint disease also imposes an individual constraint.

  - **Functional constraints** are determined by the developmental status of the individual. Functional constraints can be physiological (such as strength, cardiovascular endurance, body composition, and flexibility), cognitive (mental development, experience), or psychological (perception capacity [especially seeing, hearing, feeling], levels of motivation, confidence, fear, caution, and anxiety).

The interaction of the task, environment, and individual constraints determines the "fit" of the performer to the expected movement in the specific location. The individual constraints impose limits on what the performer is capable of doing, and these interact with the constraints from the task and the environment to influence the likelihood of success during the performance. The structural and functional constraints limit how the individual will perform the given task in the specific environment.

If a constraint is recognized as preventing goal attainment, a great way to improve performance is to make an adaptation or adaptations to create a better fit among the task, the environment, and the individual. Task constraints are overcome by adapting the rules. Environmental constraints are overcome by adapting the physical layout. And individual constraints are overcome by adapting the individual.

I revisit these possible adaptations in the "Intervening to Improve the Performance" section, later in this chapter.

Technique errors

A technique error causes failure to attain a strategic goal. A runner may be slow, a quarterback may miss the target, or a blocker in volleyball may make contact with the net.

A technique error occurs because the muscle force generated by the performer didn't create the angular velocities of the segments needed to produce the optimal impulse on the body to be moved. The easiest way to see what makes up a technique error is to explain good technique, the process of using the joint rotations to generate an external impulse.

Characterizing good technique

Successful performance depends on applying the optimal impulse in the correct direction. The force applied by muscles across the joints creates segment angular motions. The segment angular motions are used to exert an external force for a period of time on the body to be moved. Muscles must produce segment angular motions to generate an external impulse that's the right size, applied for the optimal time period, and applied in the right direction to get the desired momentum to attain the strategic goal.

The stretch-shorten cycle (see Chapter 15) is the key to muscle force production during movement. Optimal muscle force production comes through use of the stretch-shorten cycle. Actively lengthening, or stretching, a muscle immediately before it shortens during a movement lets a performer use both active and passive muscle tension to create the segment angular rotations.

Because muscle best produces force through the stretch-shorten cycle, it's critical that a performer uses the stretch-shorten cycle when moving.

Breaking down movement performance into phases

Skilled performance of almost every movement includes the following three mechanical phases of preparation, execution, and follow-through. I explain these three phases and how they relate to the stretch-shorten cycle in this section.

The preparation phase

In the *preparation phase* (sometimes called the *wind-up phase, countermovement phase,* or *pre-stretch phase*), the segments are sequentially rotated in the direction *opposite* to the motion they'll undergo when generating the external impulse.

The individual segments move "backward" in a sequence, beginning with the larger segments closer to the trunk (proximal segments) and continuing to the smaller segments farther from the trunk (distal segments).

By rotating the segments in the opposite direction, the muscles crossing the joint on the opposite side are elongated (stretched) and activated eccentrically by the stretch reflex to slow down and stop the backward angular rotation of the segments (see Chapter 15). The elongation stores passive tension in the elastic components of the muscle and creates a high level of active tension in the contractile component. (The elastic components and contractile components of muscle are explained in Chapter 15.) A high level of both active and passive muscle force develops in the elongating muscles during the preparation phase.

The preparation phase elongates muscle and prepares it to produce the muscle force to accelerate body segments during movement. The preparation phase is the "stretch" portion of the stretch-shorten cycle.

The execution phase

In the *execution phase* (sometimes called the *propulsion phase* or *acceleration phase*), the individual body segments are rotated in the direction needed to generate the external impulse on the body. The segments move "forward" during the execution phase, and the distal segment pushes or pulls to produce an external impulse.

The execution phase is when muscles use the active and passive muscle force enhanced during the preparation phase to accelerate body segments to apply the external impulse to achieve the strategic goal of a movement. The execution phase is the "shorten" portion of the stretch-shorten cycle.

The muscles previously stretched and activated during the preparation phase shorten during the execution phase, exerting both passive and active muscle force on the segments. The muscle force accelerates the angular motion of the segments, speeding them up. The faster the segments rotate, the greater the force the distal segment will exert on the body. In addition, because the segments were moved backward prior to shortening, during the execution phase the segments apply the greater force for a longer time. The greater force and the longer time of force application increase the magnitude of the external impulse applied, causing a greater change in the momentum of the body.

Muscle force applies an impulse to the segments to cause them to rotate. If the distal segment contacts, or is in contact with, another body, the angular motion of the segments allows the distal segment to apply an impulse to that body. In locomotion, the other body is the ground; the impulse applied back on the performer by the ground causes the motion of the performer's entire body to be changed. In throwing and striking activities, the other body is the object being thrown or struck; the impulse applied by the performer to the body being thrown or struck changes the motion of that body.

The follow-through phase

During the *follow-through phase* (sometimes called the *braking phase*), the body segments continue to rotate in the direction of the execution phase, but the segments are slowing down and they eventually come to rest. Depending on the direction and the speed of the angular motion, either eccentric muscle activity or gravity provides the force to slow down and stop the rotation of the segments.

Although heavily emphasized as a key to good performance, the follow-through doesn't directly affect the impulse applied to the body. If the segments are rotating quickly at the end of the execution phase, they'll continue to rotate until brought to rest, and a follow-through will be evident. If the segments are *not* moving quickly at the end of the execution phase, little or no follow-through will be exhibited by the performer. Asking the performer to "follow through" without explaining how to use the preparation and execution phases to get the segments rotating fast doesn't help. The follow-through is a *symptom* of good performance, not a *cause* of good performance.

Keying in on the mechanical phases

The external impulse created at the distal segment reflects the summed effect of all joint rotations used by the performer and the effective proximal-to-distal (larger-to-smaller) sequencing of the included segments.

For effective movement, all segments involved in the movement should go through all three mechanical phases. The phases actually overlap across the joints involved in the movement. When one joint is in the follow-through phase, other joints can be in the execution phase, and others can still be in the preparation phase.

This sequencing of the phases leads to three critical aspects of the segment or joint rotations to watch for when troubleshooting performance:

✔ **Summation of the joint forces:** Each rotating joint contributes to the external impulse. The more segments that are rotated during the performance, the greater the applied impulse by the distal segment. Conversely, to create a larger applied impulse, a performer can use additional joints in the performance. Failure to use all possible joints is a cause of a technique error because this results in a smaller applied impulse by the distal segment.

✔ **Continuity of the joint forces:** The sequencing of the segment rotations should begin with the larger, more proximal segments and flow out sequentially toward the smaller, more distal segments. The overlapping joint actions from proximal to distal should be smooth and continuous with no hesitation or pauses during both the preparation and execution phases. Failure to use smooth, continuous joint actions from larger, proximal segments to smaller, distal segments is a cause of a technique error because any pause results in a smaller applied impulse by the distal segment.

The overlapping proximal-to-distal sequencing of segment rotations means that the execution phase of a proximal segment will begin before, and contribute to, the preparation phase of the adjoining distal segment. Essentially, as the proximal segment begins rotating during its execution phase, the more distal segment continues to rotate in the opposite direction because of inertia. The opposite direction of motion of the two segments increases the elongation of the muscles crossing the joint where the segments meet. This is of benefit because the muscles stretched will pull on the distal segment during the execution phase.

✔ **Direction of impulse:** The rotations of the segments must create an external impulse that's applied in the direction of the intended momentum of the body. In locomotion, the push of the foot must be opposite to the direction of travel. In throwing or striking activities, the push on the thrown or struck object must be in the intended direction of travel.

The external impulse created at the distal segment reflects the summed effect of all joint rotations used by the performer and the effective proximal-to-distal sequencing of the included segments.

Pitching by the phases

The preparation, execution, and follow-through phases (see "Breaking down movement performance into phases," earlier in this chapter) are generally present in all movements when performed effectively. In this section, I show these three phases with pitching a baseball, an activity in which the phases are clearly evident, and then I explain some common technique errors.

Ignoring all the subtle strategy of the game, the strategic goal of a pitch in baseball is for the pitcher to throw the ball so that the batter can't hit it successfully. The mechanical objective of a pitch is for the pitcher to apply an impulse to the ball using his hand (the distal segment of his arm). The impulse applied gives the ball momentum, hopefully to go to the pitcher's intended target. The impulse applied to the ball comes from the summation of the continuous joint actions used by the pitcher during the phases of the throwing performance.

Exploiting all the phases

Figure 16-3 shows images from a video recording of an expert throwing a pitch in baseball. I refer to the shaded arm as "the throwing arm," the shaded leg as "the push-off leg," and the non-shaded leg as "the lead leg." The phases for each joint are identified below the images.

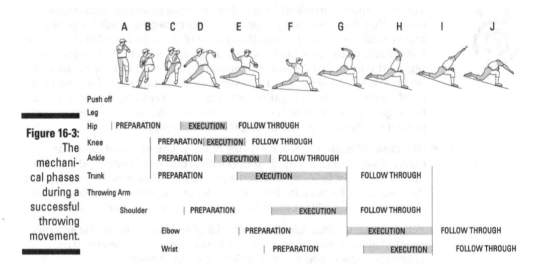

Figure 16-3: The mechanical phases during a successful throwing movement.

From A, the pitcher begins the throw by lifting the lead leg and rotating it across his body while bending the hip, knee, and ankle of the push-off leg. This starts the preparation phase for the push-off leg, stretching the muscles on the back of his hips from A to B, the front of his knee from B to C, and the back of his ankle from B to D.

From C through F, the pitcher rapidly straightens the joints of the push-off leg, beginning with the hip, then the knee, then the ankle, pushing forward to a firm planting of his lead foot. The execution phases of the joints of the push-off leg are sequential — for the hip from B to D, the knee from C to E, and the ankle from D to F. The sequencing of proximal to distal joint rotations contributes to a successful performance.

Note that from B through E, his trunk (upper body) is in the preparation phase, as the extending legs rotate from under the trunk. The forward rotation of the legs and the backward rotation of the trunk stretch the muscles on the front and side of his abdomen. The execution phase of the trunk begins as the heel of the lead leg contacts the ground (E) and the stretched muscles shorten and rotate his trunk in the direction of the throw. The ground pushes back on the pitcher, creating a torque to supplement the muscle force rotating the pitcher's trunk in the direction of the intended throw.

The phases of the throwing arm lag behind the phases of the leg and trunk, allowing the forces produced by the leg and trunk to supplement those produced by the arm. Beginning at C, the throwing arm rotates away from and backward relative to the trunk. When the trunk begins its execution phase at E, starting to rotate in the direction of the throw, the arm continues to move in the opposite direction; these opposite motions enhance the preparation phase at the shoulder, elbow, and wrist as the trunk essentially moves out from under the arm.

The execution phases of the arm also show the proximal-to-distal sequencing of the joint rotations: The shoulder begins (F) before the elbow (G), which begins before the wrist (H). The ball is released just after the wrist begins its execution phase. The execution phases of all the segments allow the distal segment, the hand, to apply a large external impulse to the ball from D to I, increasing its momentum to a very high value.

Because the arm is rotating while the ball is in the pitcher's hand, the instant when the pitcher releases the ball is critical. As the arm rotates, the ball in the pitcher's hand has a tangential velocity (see Chapter 5) that's being changed by the pull of the pitcher's hand. At release, the ball becomes a projectile and will follow a parabolic path (projectile motion and the parabolic path are also explained in Chapter 5). The pitcher must release the ball at the instant that its tangential velocity will make it follow a parabolic path toward his targeted location. After the ball is released, the pitcher can no longer affect the motion of the ball. The ball is a projectile from the instant of release until it's hit by the batter or caught by the catcher, and the motion of a projectile is affected only by gravity and air resistance (the effect of air resistance on a baseball is explained in Chapter 11).

Figures 16-1 and 16-2 show static "mental images" associated with good throwing. The "leading elbow" position of Figure 16-1 is the same as H in Figure 16-3. The elbow position in H occurs *because* of all the joint actions between A and H. Without those joint rotations, the elbow position won't be evident, regardless of whether you ask the pitcher to "lead with the elbow."

Similarly, the follow-through position (J) of Figure 16-3 is the same as the "mental image" of Figure 16-2. The follow-through position in J occurs *because* the joint actions between A and I caused the segments to rotate quickly. Without those joint rotations, the follow-through won't be evident, regardless of whether you tell the pitcher to "follow through."

In addition, although release of the ball is typically thought to begin the follow-through phase, you can see in the images that the follow-through is not a single phase but occurs sequentially across all the joints used in the movement. The follow-through for the hip begins in D and ends in F, while the follow-through for the wrist begins in I and ends in J.

Limiting some of the phases

Although throwing is an important part of many different activities and sports, not everyone throws like the expert described in the previous section. Developing effective throwing technique usually enhances participation in a sport based on throwing.

Earlier, I refer to the common cues of "lead with the elbow" and "follow through" used when working with a performer to improve throwing, and I point out that these are *symptoms* of performance, not the *cause* of performance. In this section, I explain more relevant cues based on improving the preparation and execution phases in more joints — factors that are important to success in throwing (and causing the follow-through to be present in the performance).

Developing the throwing pattern

The process of throwing a ball has been studied as a model of how complicated movements are learned by a performer. The research has identified a four-stage sequence of throwing development, shown in Figure 16-4. Identifying a performer's current stage in the sequence allows for the instruction of the next stage, which has been shown to be an effective way to develop throwing ability.

(1) (2)　　　(1) (2)　　　(1) (2)　　　(1) (2)

Figure 16-4: The developmental stages of the throwing movement.

Stage 1	Stage 2	Stage 3	Stage 4
1) Both feet stationary	1) Feet stationary	1) Forward step with leg on same side as throwing arm	1) Forward step with leg opposite to throwing arm
2) Body face direction of throw	2) Limited trunk rotation	2) Arm raised up and behind the trunk at the shoulder	2) Greater trunk rotation, separate from arm action
3) No or minimal trunk rotation	3) Upper arm moves horizontally, still mostly elbow and wrist motion with no preparation phase	3) Trunk rotates and bends forward at same time as the shoulder, elbow, and wrist motion	3) Arm lags behind trunk rotation
4) Primarily elbow and wrist extension, with no preparation phase			4) Shoulder, elbow, and wrist rotate in proximal to distal sequence

The development of throwing from Stage 1 through Stage 4 basically involves incorporating more joints into the movement, using the preparation and execution phases at each joint, and sequencing the joint rotations from proximal to distal:

- ✔ **Stage 1:** Elbow extension and wrist flexion alone are used to apply impulse to the ball, and the preparation phase is not used at either joint.

- ✔ **Stage 2:** Trunk rotation is included in the performance, but the preparation phase for the trunk is minimal. There is limited proximal-to-distal sequencing of the joint actions because the upper arm moves horizontally along with the forward trunk rotation in a single execution phase, and the elbow and wrist joints don't show a preparation phase.

- ✔ **Stage 3:** Leg action is added to the throwing movement, but the lead leg is on the same side as the throwing arm, and there is minimal preparation phase for the ankle, knee, and hip joints of the push-off leg. The trunk rotates opposite to the direction of the throw in a limited preparation phase. The arm remains the main contributor to the throw, but it's raised up behind the shoulder by the performer and not because the trunk moves out from under it. The rotations of the joints of the arm don't lag behind the trunk during either the preparation phase or the execution phase.

- ✔ **Stage 4:** The leg on the same side as the throwing arm becomes the push-off leg, and the opposite leg is the lead leg. The proximal-to-distal sequencing of the individual preparation and execution phases across the joints of the leg, the trunk, and the arms is used by the performer. The joint rotations and sequencing in Stage 4 are essentially those demonstrated by the expert pitcher described in the previous section, but they differ in how fast and how far the rotations occur.

This example, specific to a baseball pitch, shows that developing a "good level" of performance technique involves incorporating the fundamental principles of summation of joint forces, continuity of joint forces, and direction of joint forces to apply the optimal impulse for a given situation. A fundamental "good level" of technique is similarly attained for all movements, because momentum is developed by applying impulse. Incorporating the preparation and execution phases sequentially across all the joints that can contribute to the movement enhances the impulse applied to develop momentum and causes a "good level" of performance.

I explain the basics of intervening to improve performance in the next section.

Intervening to Improve the Performance

Intervening involves taking steps to improve performance. The intervention can be focused on overcoming constraints or giving feedback related to technique with the goal of improving the impulse produced by the performer.

Adapting the constraints on throwing performance

Constraints are the task, environment, and individual factors that can limit performance. In throwing, numerous constraints in each category can affect performance by the beginning throwers of Stage 1 through to the expert performers in Stage 4. Here's an example of each category of constraint as related to throwing performance:

✔ **Task constraints:** Although the dimensions of the field, bat, and gloves used by 10-year-olds differ from those of professional players, the rules of organized baseball require that the baseball used by 10-year-olds is the same dimensions and weight as the baseball used by pro players. An average 10-year-old male is 4'7" (140 cm), while the average professional pitcher is 6'2" (188 cm). The regulation baseball has a circumference of 9 to 9¼ inches (22.9 to 23.5 cm) and weighs 5 to 5.25 ounces (1.4 to 1.5 Newtons).

The required baseball size represents a task constraint on the throwing performance of the 10-year-olds. With smaller hands and less strength than their professional counterparts, the average 10-year-old is less able to exert and control the impulse applied to the baseball while throwing. Changing the rules to allow use of a smaller, lighter ball would allow the 10-year-old pitchers to throw faster, maintain better control over the ball (because the instant of release is so critical to successful throwing, as explained earlier), and use the different grips on the ball to experiment with a variety of different pitches (fastball, slider, curveball, and so on).

Of course, the question of whether 10-year-olds should actually pitch in baseball games is separate from the question of scaling down the ball. In some youth leagues, pitching machines are used to reduce the loading on the arms of young players and to provide the batters with more consistent pitches (and less fear of being hit by a wild ball) as the batting movement is developed.

✔ **Environmental constraints:** The baseball field used by younger players in organized leagues has scaled-down dimensions, reflecting the shorter height and reduced strength of the players. The shorter dimensions increase the chance of success when throwing a ball and are an example of adapting the environment to match the performer.

One environmental constraint at both the 10-year-old and professional levels of baseball is the presence of spectators. Many professional players thrive on the roar of the crowd, but the anxiety of performing well for others, especially parents, adversely affects many youth players (and it does for some professional players, too). Throwing is a complicated movement, and for a younger player still developing an effective throwing movement, failure to perform well in front of others can lead to dropping out of organized baseball. Some organized sport leagues actually prohibit parents from attending games and practices, to allow kids to play for the sake of playing.

✔ **Individual constraints:** The shorter height and reduced strength of a 10-year-old compared to a professional player represent individual constraints, and many rule makers think kids should get used to the larger baseball "because that's how baseball is played." However, a 10-year-old could be taller and stronger than average for his age but still be shorter and weaker than the older professional player and be disadvantaged by the size of the ball.

This is why it's important to adapt the task and environment to fit the child instead of waiting for normal growth and development to adapt to the task and environmental constraints. Actually, adapting the task and the environment to a recreational performer of any age or developmental status is a great way to enhance enjoyment, and continued participation, in any movement.

When the task and environment can't be adapted, for example in elite-level competitive activities, the performer must overcome the structural and functional constraints to perform as and where required. Training to improve strength, endurance, and flexibility are examples of adapting the performer to overcome individual constraints.

Refining technique

The process refers to how the performer produces impulse. Errors in producing impulse are the cause of poor technique. Feedback to a performer to improve performance must be specific to the production of impulse used to develop the required momentum of the body.

Many of the static positions identified as "coaching cues" and used as the basis of intervention for a less-than-satisfying performance are not the causes of impulse production. A common intervention is for the movement analyst to pose in a desired position to demonstrate "proper technique." For example, in throwing, the analyst may try to show the "lead-with-the-elbow position," although rarely can anyone with a healthy elbow or shoulder simply stand and demonstrate the arm position in Figure 16-1. Another approach is to tell the learner to watch someone who does match the "mental image," as the analyst points out to "Watch his elbow — look at how it leads the arm" or "Watch how she follows through." But as stated earlier, a focus on static positions ignores how the body gets into the desired position of the mental image.

The critical phases of a performance are the preparation and execution phases. When used in a performance, both active and passive tension from the muscles pulls on the segments. The pull of the muscles causes the angular motion of the segments, and the angular motion of the segments produces the impulse at the distal segment.

Intervention based on producing impulse should focus on the following points:

- ✔ **All joints used in a movement should first move "backward" (preparation phase) before moving "forward" in the direction of the desired impulse (execution phase) to take advantage of the stretch-shorten cycle.** Incorporating the stretch-shorten cycle improves the muscle torque applied to segments to make them rotate quickly and results in a larger impulse.

- ✔ **Use all joints that can contribute to the performance.** Leaving out a joint or joints reduces the applied impulse.

- ✔ **Use all joints sequentially, from the joints of the larger segments (typically, the proximal segments) to the joints of the smaller segments (the distal segments).** This allows optimal force to be produced by the larger muscles that cross the joints and control the more proximal segments.

- ✔ **Feedback to the performer should focus on teaching the movement in parts.** The cues should be directed toward the larger, more proximal segments. When these segments move correctly, the more distal segments will lag behind because of inertia, and the desired "mental image" of good technique will be seen in the performance.

Chapter 17

Putting a Number on Performance: Quantitative Analysis

*I*n biomechanics, a *quantitative analysis* involves measuring some aspect of performance to give it a numeric value. A quantitative analysis tells us "how much" and can be as simple as counting the number of laps completed by a runner, timing how long it takes a football player to run 40 meters (m), or measuring the distance of a kick. A quantitative analysis can also involve the use of sophisticated instruments to measure kinematics, kinetics, or muscle activity.

The purpose of a quantitative analysis is to provide a better understanding of how a body moves than can be gained without the measurements. Assigning a numeric value makes it easier to compare different performances. Instead of just saying a patient recovering from knee surgery is now faster, it can be said that forward velocity increased by 0.2 meters per second (m/s) or an increase of 6 percent compared to the last measurement.

A quantitative analysis involves data collection and data processing. In this chapter, I introduce systems for quantifying *kinematics* (movement patterns), *kinetics* (contact forces during movement), and muscle activity.

Converting Continuous Data to Numbers

To measure something means to assign a value to it at a particular instant in time. Each measurement is known as a *sample,* and how frequently the measures are taken is called the *sampling rate* or *sampling frequency.* Sampling frequency is reported in Hertz (Hz). If something is measured 100 times in a second, the sampling rate is 100 Hz, and the time between each sample is calculated as 1 second/Hz; for a sampling frequency of 100 Hz, the time between samples is 1 second/100 or 0.01 second.

In movements of interest in biomechanics, variables such as position, force, and muscle activity vary continuously. A variable that changes continuously over time is called an *analog signal.* Sampling an analog signal gives a *digital* (numeric) value of the signal at an instant in time. The process of sampling an analog signal to create a digital value is known as an *analog to digital conversion,* or A/D conversion for short. Because it's a sampling frequency, the process of A/D conversion is reported in Hertz.

To gain an understanding of a movement, the variables of interest must be sampled multiple times during performance to create a time history of the variable. A *time history* consists of the multiple individual values of the variable sampled, showing how it changes during the performance.

A critical idea when a variable is sampled is to set a sampling frequency high enough that the set of discrete values in the time history gives a true representation of how the variable changes throughout the performance. If the sampling frequency is too low, the time delay between the data points means that a change in value of importance may be missed, and the time history doesn't represent the performance. If the sampling frequency is high enough, with a short time between samples, the time history can be considered as a set of instantaneous values of the variable during the performance.

I refer to sampling frequency in the following sections as I describe technology used to quantify kinematics, kinetics, and muscle activity, and I give examples of the time history of these variables.

Measuring Kinematics: Motion-Capture Systems

Kinematics is the branch of mechanics that describes motion. Typical variables in kinematic analysis include position, velocity, and acceleration. (See Chapter 5 for a description of linear kinematics and Chapter 9 for a description of angular kinematics.)

The most common instrumentation used to measure linear and angular kinematics is a motion-capture system. Basically, a motion-capture system consists of a "camera" system, to provide the A/D conversion of the continuous movement into a series of individual images, and software installed on a computer, to process the individual images into the desired kinematic variables. In this section, I describe the basic collection and data processing of a motion-capture system.

Collecting kinematic data

Today, most images aren't recorded with a traditional camera like the one your parents used to record the family picnic. No visible image of the performer is recorded. Instead, each camera emits an infrared light that's reflected off markers attached to the performer and back to the lens of the camera. The light reflected back from the markers is all that's needed for processing the kinematic data from the performer.

Figure 17-1 shows a typical camera setup, consisting of six cameras aimed toward where a movement of interest will be performed. The target area is called the *capture volume,* and the volume must be set large enough so that all the movement is performed within this volume. The number of cameras used is often more than six to capture a larger volume and to ensure that at least two cameras track each marker on the performer at all times during the performance. The sampling frequency of the cameras is typically set to between 100 Hz and 500 Hz. A higher frequency is used with faster movement. Each recorded image is called a *frame,* a holdover from when actual cameras were used to record the performer. At a sampling frequency of 200 Hz, 200 frames are imaged each second.

Figure 17-1:
A typical laboratory set up to collect kinematic data.

The reflective markers are typically 1 cm balls covered with reflective tape. The number and placement of the markers depends on the activity being analyzed. A set of 13 markers used to record the motion of the legs during locomotion or landing is shown in Figure 17-2. Each marker is secured to the skin of the participant with double-sided tape in locations that can be used to create a link segment model of the pelvis, thighs, lower legs, and feet. (For more on the link segment model, turn to Chapter 8.)

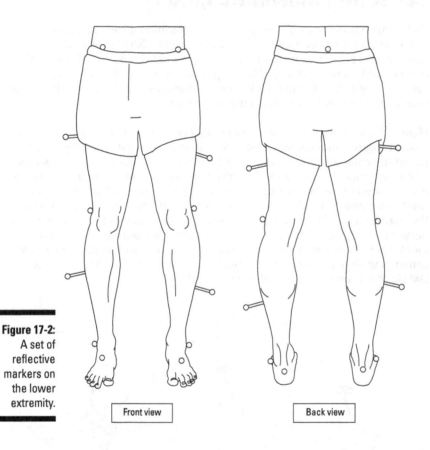

Figure 17-2:
A set of reflective markers on the lower extremity.

Front view   Back view

Processing kinematic data

A critical step to obtaining accurate kinematic data is calibrating the camera setup before recording the performance. This means putting a three-dimensional grid of reflective markers spaced a known distance apart into the capture volume and recording them. The grid sets an origin for a coordinate system within the capture volume used for measuring performance. Generally, the camera setup is calibrated before each data-collection session.

During performance, the cameras track the markers, and the multiple images are stored in the computer. Special software is used to assign a pair of *x*- and *y*-coordinates to each marker recorded in each image by each camera, known as *digitizing the coordinates*. The software then uses a process called *direct linear transformation* (DLT) to combine the two-dimensional coordinates of each marker from each camera into a single set of three-dimensional coordinates representing the position of each marker in the calibrated capture volume.

The calibrated coordinates are used to generate the link segment model of the body. An example of the link segment model of the right leg created from digitized coordinates during a single stride of walking is shown in Figure 17-3. The positions of the hip, knee, and ankle joint centers are identified from the tracked markers. Joining the hip and knee defines the thigh, and joining the knee and ankle defines the lower leg, or *shank*.

Figure 17-3:
The link segment model of the right leg.

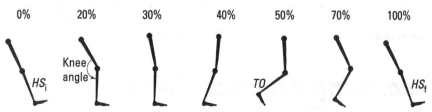

A stride of walking contains a ground contact phase, between the initial heel strike (*HS_i*) and toe-off (*TO*), and a swing phase, between *TO* and the final heel strike (*HS_f*).

Pretty much any type of kinematic data, including linear and angular position, velocity, and acceleration values, can be calculated from the coordinates of the link segment model in each recorded frame. Graphing the data calculated from individual frames provides a time history of the variable during the performance.

Figure 17-4 shows the time-history plot of the knee joint angle during the stride. The knee joint angle is measured as the relative angle between the thigh and the shank, as indicated in Figure 17-3. (See Chapter 9 for more on how the relative angle of the knee joint is calculated from the thigh and shank segments.) The equations for calculating joint angles are included in the motion-capture analysis software.

From the time history of a variable, specific values of interest can be measured. For example, knee angle at HS_i (172 degrees) and maximum knee flexion during the ground contact phase (158 degrees at 14 percent of the gait cycle) are frequently reported for gait, and the range of motion, or angular displacement, during this joint action is calculated as $\theta_f - \theta_i = 158° - 172° = -14°$, with the negative sign indicating joint flexion. (See Chapter 9 for more on calculating angular displacement.)

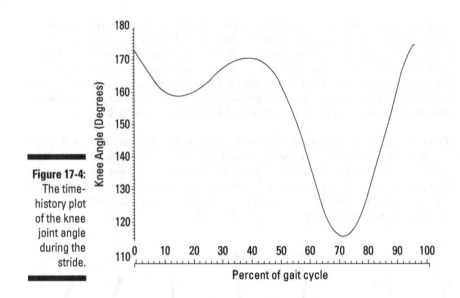

Measuring Kinetics: Force Platform Systems

Kinetics is the branch of mechanics describing the forces that act on a body to cause a change in motion. Typical variables quantified in kinetics include the forces that cause a change in linear motion and the turning effect of a force, called *torque*, which causes a change in angular motion. (See Chapters 4 and 6 for more on force, and Chapter 8 for more on torque.)

The most commonly used instrumentation to measure force is a *force platform system,* which consists of the force platform and software installed on a computer to record and process the force data. In this section, I describe the basic collection and processing of force using a force platform system.

Collecting kinetic data

A force platform measures the contact force between a body and the platform. The top surface of the platform is the measuring surface. Figure 17-5 shows a force platform installed with the measuring surface flush with the floor in a biomechanics lab. Multiple force sensors (technically called *transducers*) are attached to the measuring surface inside the force plate. The sensors are aligned to measure three components of the force applied to the platform surface: the normal contact force in the vertical direction, and the friction force in the antero-posterior and medio-lateral directions. (Normal force and friction force are explained in Chapter 4.)

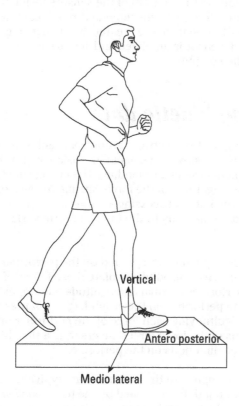

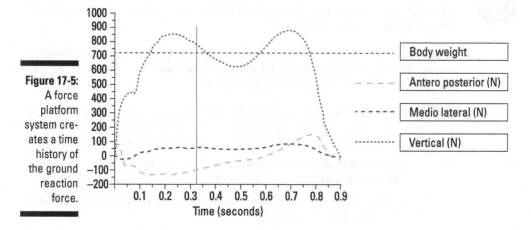

Figure 17-5: A force platform system creates a time history of the ground reaction force.

Forces can't be seen, but the effect of the force can be detected. When a body is in contact with the measuring surface, like the foot of the participant walking in Figure 17-5, the force from the body causes a non-visible deformation (or *strain,* described in Chapter 12) of the sensors inside the platform. The strain changes an electric current running through each sensor. The greater the deformation of the sensor, the greater the change in electric current. The current from each sensor is amplified and sent to be sampled and processed by software in the computer.

Processing kinetic data

The software samples the electric current from each sensor, with typical rates about 100 Hz for slow movements like balance or as high as 2,000 Hz for faster movements like gait or landing. The change in electric current is calibrated to the magnitude of the force causing the sensor to deform, so the output from the initial processing are measures of the components of the force applied by the body to the force platform surface at each instant of sampling.

Because the force applied by the ground on the performer is what's of interest, Newton's third law is used to calculate these forces: The force of the ground on the performer is equal in magnitude and opposite in direction to the force of the performer on the ground. (See Chapter 6 for more on Newton's laws.) Multiplying each force of the performer on the platform by –1 reverses the signs of the force direction, creating a time history of the ground forces in the three-directions on the performer.

The ground forces applied to the performer are typically called *ground reaction forces* (GRFs), but don't be misled by the term *reaction* — there's no contact force without the two bodies touching, and one force doesn't cause the other force as a reaction (see Chapter 4 for more on this). But because GRF is the common term, I use it, too.

A time history of the three components of the GRF during the ground contact phase of the right leg during walking is shown in Figure 17-5. Here are important features of each of the GRF components:

✔ **Vertical GRF:** The vertical GRF is the largest of the three components and acts only upward (in the + direction) on the walker. There are two phases when the vertical GRF is greater than the individual's body weight:

- The first phase creates a net positive force to decrease and stop the downward velocity of the walker.

- The second phase creates another net positive force to increase the upward velocity of the walker.

✔ **Anterior-posterior GRF:** The anterior-posterior GRF shows an initial phase of negative force, indicating when the ground pushes back on the walker, followed by a phase of positive force, when the ground pushes forward on the walker. The negative phase is called the *braking phase,* because the negative impulse decreases, or slows down, the forward velocity of the walker. The positive phase is called the *propulsive phase,* because the positive impulse increases, or speeds up, the forward velocity of the walker.

✔ **Medio-lateral GRF:** The medio-lateral GRF is typically the smallest component of the contact force when walking. Whereas the vertical and anterior-posterior forces mostly affect the motion of the entire body, the medio-lateral GRF is very much affected by the placement of the foot relative to the midline of the performer's body and the motion of the foot while on the ground.

From the time history of each GRF component, specific values of interest can be measured — for example, the peak values of the two phases when the vertical GRF is above body weight and the peak of the braking and propulsive forces are frequently reported. Calculating and comparing the impulse of the braking and propulsive impulses (see Chapter 6) of the anterior-posterior GRF indicates whether the performer speeds up, slows down, or stays at the same forward velocity between the start and end of a ground contact phase.

Recording Muscle Activity: Electromyography

The central nervous system controls muscle tension by communicating low-voltage (millivolt) signals called *action potentials* (see Chapter 14). Electromyography (EMG) is the study of muscle activity and involves recording and quantifying the action potentials that activate a muscle.

The most common instrumentation used to measure muscle activity includes a sensor and amplifier, used to detect and amplify the voltage of the action potentials, and software installed on a computer for sampling, recording, and processing the collected data, called an *electromyogram.* In this section, I describe the basics of collecting and processing the surface electromyogram.

The sensors used for EMG are often called *electrodes,* but the term *electrode* suggests that an electric shock may be possible. (People have watched a lot of horror movies.) Using the term *sensor* is less intimidating and removes this unwarranted fear.

Collecting the electromyogram

The most common method to detect the electromyogram is *surface EMG*, which uses a sensor with two pickups and a small amplifier (called a *bipolar sensor*) secured to the surface of the skin over a muscle of interest. Surface EMG is usually used on muscles just under the skin. Using a bipolar sensor and a ground attached to a bony prominence on the body obtains a cleaner signal of the muscle activity.

In-dwelling sensors can also be used for EMG. Using a needle, the sensors are injected directly into the muscle fibers to record the action potentials. In-dwelling sensors are used to measure muscles deeper under the skin surface or to measure the action potentials from individual motor units.

At any instant there are multiple action potentials present on fibers of different motor units of an individual muscle. This is shown in Figure 17-6, where two motor units (α_A and α_B) innervate the fibers (numbered 1 through 5) of a muscle. The size and shape of the action potential reaching the sensor from each of the fibers is shown in Figure 17-6 and depends on a variety of factors, including how deep the fiber is below the sensor, blood flow in the fiber, and the amount of body fat between the sensor and the muscle fiber. The action potential has a positive and negative phase, reflecting where it's recorded as it goes along the fiber away from the motor end plate. The surface EMG at any instant is actually the sum of the multiple action potentials detected by the sensor, indicated in Figure 17-6 as ΣMUAP (for sum of the motor unit action potentials). The ΣMUAP detected at the sensor is sent to be sampled and processed by software in the computer. EMG is typically sampled at 1,000 Hz.

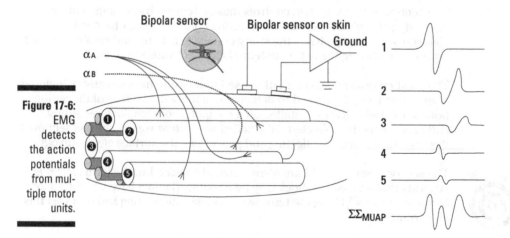

Figure 17-6: EMG detects the action potentials from multiple motor units.

Processing the electromyogram

A typical recorded electromyogram, in this case from the biceps brachii muscle crossing the elbow joint on the front of the arm, is shown in the top graph of Figure 17-7. This raw, or unprocessed EMG, was recorded while asking the participant to twice try to flex (bend) the elbow while the joint is restrained from moving (isometric activity of the biceps, as explained in Chapter 15).

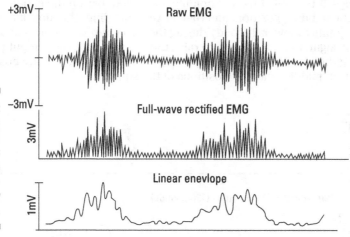

Figure 17-7: Processing EMG involves rectifying and smoothing the raw data.

The raw EMG data shows sharp negative and positive peaks fluctuating around a baseline value. Quantifying raw EMG is limited, but the start and end of muscle activity can be identified, indicated with the short vertical lines, and these values are used to measure the duration of muscle activity.

Processing the raw EMG first involves *rectifying the signal,* which means to take the absolute value of each recorded data point. All data points are positive in the full-wave rectified EMG shown in the middle of Figure 17-7.

The sharp peaks and valleys in the EMG are smoothed out in the next step. *Smoothing* is basically a process of taking a "moving average" of the rectified values, with each smoothed point calculated as the average of rectified points before it and after it. Smoothing the rectified electromyogram creates a linear envelope, as shown in the bottom of Figure 17-7. The shape of the linear envelope closely resembles the shape of the muscle tension produced by the muscle from which the electromyogram is measured. Higher values of the EMG linear envelope correspond to higher values of muscle tension.

EMG does not directly measure muscle tension. Also, without sophisticated analysis, EMG does not indicate if the increase in muscle activity is from motor unit recruitment or rate coding, the two processes used by the central nervous system to increase the tension from a muscle. (See Chapter 14 for an explanation of these two processes.) But the interpretation of increased tension with increased muscle activity is correct.

EMG is very useful when it's used with a motion-capture system to measure muscle activity and joint position at the same time, with the data sampled simultaneously. The squat is an exercise used to develop muscle strength of the legs. It typically involves holding a weighted bar on the shoulders in an upright standing position, and then *flexing* (bending) the hips, knees, and ankle joints to lower the body during the descent phase, followed by *extending* (straightening) the joints to raise the body back to the upright position during the ascent phase. The top of Figure 17-8 shows the body position at the start, middle, and end positions of the squat.

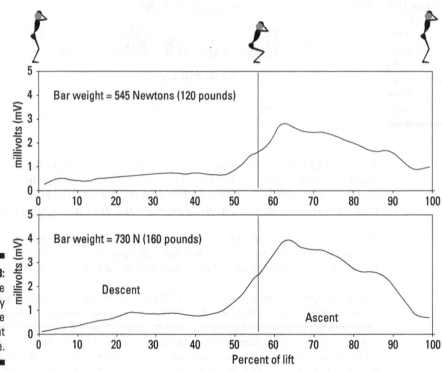

Figure 17-8: Muscle activity during the squat exercise.

The vastus lateralis muscle crosses the front of the knee joint. As the knee joint flexes, the vastus lateralis gets longer. As the knee joint extends, the vastus lateralis gets shorter. The start and end of knee flexion, and the start and end of knee extension during the squat, can be identified from the knee position time history output from the motion-capture system, as described earlier.

The bottom portion of Figure 17-8 shows two linear envelopes of the vastus lateralis when a lifter performed the squat, the top with a bar weighing 545 N and the bottom with a bar weighing 730 N. Because the start and end of the ascent and descent phase were identified from the joint position data collected by the motion-capture system, the linear envelope can be divided into the descent and ascent phases.

Eyeballing the linear envelopes for the lifts shows two things:

- ✔ **Muscle activity is greater during both phases of the lift with heavier weight, as would be expected.** More muscle activity is necessary to perform a lift with heavier weight.

- ✔ **Muscle activity is greater during the ascent phase than during the descent phase.** During the descent phase, the vastus lateralis is active and getting longer, which is *eccentric muscle activity.* During the ascent phase, the vastus lateralis is active and getting shorter, which is *concentric muscle activity.* Because a muscle can produce more force eccentrically than it can concentrically, the level of vastus lateralis activity is lower during the descent phase than it is during the ascent phase to produce the force to complete the phase. (See Chapter 15 for more on eccentric and concentric muscle activity.)

Integrated EMG (iEMG) is the area under the EMG curve. Calculating iEMG quantifies the muscle activity during each phase. The iEMG can be compared between phases of the squat to show numerically the difference in eccentric and concentric muscle activity.

The iEMG is calculated using the individual data points from the linear envelope. With a sampling rate of 1,000 Hz, the time between each data point is 0.001 seconds. Adding up the value in millivolts of each EMG data point and multiplying this sum by 0.001 calculates the area under the linear envelope in mV · s, the units of EMG (millivolts) multiplied by the units of time (seconds). Software performs this calculation very quickly.

Chapter 18

Furthering Biomechanics: Research Applications

*R*esearch is a process of systematically increasing our knowledge about something. Research basically stems from curiosity and involves exploring how and why things work as they do. The process involves posing a question, making careful observations and precise measurement to seek an answer, and writing up and communicating the results for others to critically evaluate and consider as an answer to the question posed. Accepted answers represent our knowledge of the universe. However, knowledge is not stagnant. The purpose of research is to refine, and sometimes challenge, our currently accepted answers.

Research using biomechanics is an integral part of understanding how and why living things move as they do, and get injured as they do. The questions investigated by biomechanists are very diverse, reflecting the different ways biomechanics can be utilized to understand human performance and limitations. In this chapter, I provide an overview of some applications of biomechanics in past and current research from a variety of different fields.

Exercising in Space

Although images of astronauts floating in the space station are intriguing, the zero gravitational load responsible for the weightless environment of space flight presents a long-term challenge for bone and muscle health. Without the need to work against the constant gravitational load, bone density and muscle strength decrease. The lower bone mineral density and muscle strength create significant problems for an astronaut after returning to earth following extended space flight.

The challenge is to design exercise equipment for use in space allowing maintenance of muscle strength and bone density. Obviously, dumbbells and barbells can't be used, because the bars and plates are weightless in space, just like the astronaut! Instead, exercise machines and treadmills are designed using rubber cables. Working against the resistance of the cables is intended to provide an external load to the bones and muscles when used during exercise.

However, setting up an exercise area in the space station is problematic. Because the entire space station is weightless, an exercise machine or treadmill simply secured to the floor or walls of the space station creates significant vibration throughout the space station when used by an exercising astronaut. The vibration can adversely affect other experiments being conducted in the space station designed to take advantage of zero gravity and can damage the space station itself.

The solution for the first treadmill used in space was to design it to float within the space station, held in place by a gyroscope. This system, called the Treadmill with Vibration Isolation System (TVIS), created a beneficial problem. The running astronaut had to contend with the wobble of the treadmill as it floated in place, but this actually served as a form of sensory training in addition to the cardiovascular and bone benefits of the exercise. A newer treadmill, with a wider running belt, replaces the gyroscope with springs to reduce the vibration.

Designing a treadmill on earth for use in space required simulating zero gravity. Creative researchers mounted a treadmill equipped with a force platform (for more on force platforms, see Chapter 17) on a wall. The runner was suspended horizontally by rubber cables to run on the treadmill. The cables provided enough upward force to make the runner "weightless." Additional rubber straps attached at the shoulders and waist keep the runner on the treadmill. The force platform was used to measure the ground reaction forces as the runner ran on the treadmill, allowing for adjustment of the shoulder and waist cables to create a load similar to running on earth.

Repairing the Anterior Cruciate Ligament

People frequently sprain *ligaments,* the tissue joining bone to bone and providing support to the joint. (Injury to ligaments is explained in Chapter 13.) The knee joint is one of the most commonly injured joints during physical activity, and a third-degree sprain, or complete tear of the anterior cruciate ligament (ACL) — which crosses the knee joint from the back of the *femur* (thigh bone) to the front of the *tibia* (shin bone), as shown in Figure 18-1 — is one of the most debilitating injuries.

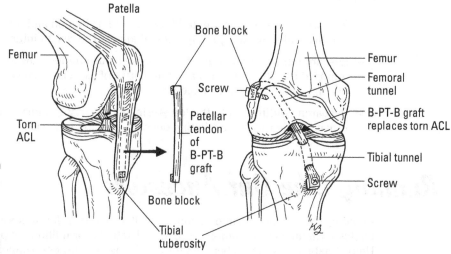

Figure 18-1:
Surgical repair of the ACL.

A torn ACL causes a lot of knee instability, limiting performance and increasing the risk of degenerative joint disease (also explained in Chapter 13). Until the mid-1980s, a torn ACL was a career-threatening injury for an athlete.

The improved surgical procedures for ACL repair provide a great example of the contributions of biomechanics research. Before 1980, surgery involved opening up the knee joint capsule and sewing the two ends of the torn ACL together. The leg was immobilized in a thigh-to-ankle cast until the ACL healed. After removing the cast, a long rehabilitation was needed to restore the muscle strength, joint flexibility, and bone mineral density lost because of the lack of imposed stress while immobilized. Even then, performance level was never completely recovered.

The improved surgical repair was developed to avoid the negative effects of immobilization and to take advantage of how strongly bone can heal. As shown in Figure 18-1, the technique involved using a graft from the patellar tendon, removed with a piece of bone from the tibia and the *patella* (knee cap) at each end. The bone–patellar tendon–bone (B-PT-B) graft was then surgically attached within the joint to the femur and to the tibia to replace the ACL. The joint was not immobilized. Flexibility exercises and earlier walking using crutches reduced the loss of flexibility and muscle strength, and the imposed stress from activity increased the strength of the bone at the graft sites in accordance with Wolff's law (explained in Chapter 13).

Since the B-PT-B technique became popular in the early 1980s, continued research has refined both the surgical procedure and the rehabilitation program:

- ✔ Researchers have investigated how the ACL is loaded and injured during activity and how knee joint alignment affects the risk of injury.
- ✔ Research into the rehabilitation programs has studied how loading imposed by different types and intensities of exercise affects graft healing and joint performance after recovery.
- ✔ Research continues into the long-term effects of ACL injury and surgical repair on the risk of developing degenerative joint disease.

Running Like Our Ancestors

Running is a common form of locomotion. It's used in different sports, and by itself, running is a common recreational activity for millions of people. Unfortunately, despite its benefits for cardiovascular fitness, runners continue to suffer a variety of overuse injuries related to the repetitive loading imposed with each ground contact.

Running has been the focus of considerable research in biomechanics. A lot of the early research consisted of a descriptive kinematic analysis of running, using film recordings of runners and focusing on the range of angular motion at the ankle, knee, and hip joints and the coordinated timing of the joint actions.

As force platforms (described in Chapter 17) became widely used in biomechanics, research became focused on the force-time history of the ground reaction forces during the *support phase* of running, the time between when the foot contacts the ground and when the toe leaves the ground.

Two typical vertical ground reaction force (vGRF) curves during running are shown in Figure 18-2. One is from a *rearfoot striker,* a runner who lands on the outside of the heel of the foot; this is a common foot-strike pattern for individuals wearing shoes. The other is from a *forefoot striker,* a runner who lands on the ball of the foot; some runners use this pattern while wearing shoes, and it's commonly the foot-strike pattern of a barefoot runner.

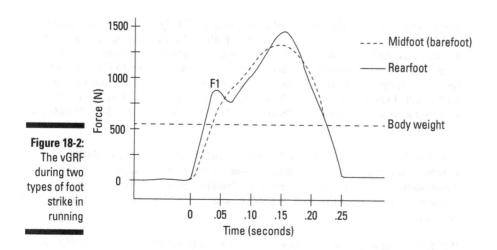

Figure 18-2:
The vGRF
during two
types of foot
strike in
running

The most striking difference between the two vGRF curves is the "spike" (identified as F1) in the curve of the rearfoot striker. F1 occurs very early during ground contact, typically within 10 to 50 milliseconds (thousandths of a second), as the heel of the foot contacts the force platform. Because it's a high force applied in a very brief period of time, F1 was implicated and investigated as a possible cause of running injuries.

Considerable early work in running research focused on designing shoes with thick cushioning heels to reduce or even eliminate F1 from the vGRF. However, this approach highlighted a problem with designing shoes based on only one aspect of measured biomechanics. The soft raised heel of a highly cushioned shoe increased the inward roll of the foot during ground contact, an action called *pronation*. The increased pronation disrupts the coordinated action of the joints of the leg, especially between the ankle and knee joints.

Current shoe designs attempt to provide cushioning while not disrupting pronation, a shoe design feature called *rearfoot control*. Pronation can be accurately measured using the high-speed motion-capture systems in a biomechanics lab (motion-capture systems are described in Chapter 17). Measuring the vGRF and pronation at the same time is useful in designing running shoes to provide both cushioning and rearfoot control.

In recent years, a completely different approach to running and footwear has been investigated. A group of clinical biomechanists suggested that running shoes, which are intended to provide a safe interface between the runner and the ground, may actually *increase* the risk of injury.

Noting that the widespread use of shoes is a relatively recent occurrence in the evolution of man, these researchers proposed that running shoes create an artificial environment for the foot. The sole of the foot contains many sensory receptors sensitive to pressure. The stimulation from the sensory receptors helps the bones, ligaments, and muscles of the foot adapt to provide natural shock absorption during all forms of gait, including running. Wearing a shoe reduces the sensitivity of the foot, leading to a loss of natural shock absorption.

The development of minimalist footwear, and the increased popularity of barefoot running, was an outcome of this research. Minimalist footwear is essentially a running sandal; the sandal consists primarily of a tough outer surface on the sole of the foot to protect against cuts and abrasions from the running surface.

When wearing a minimalist shoe, runners tend to adopt a forefoot-strike running style (similar to the style of running typically found in barefoot runners). As I mention earlier, in a forefoot strike, the foot contacts the ground in a toes-down position (called *plantarflexed*). As shown in Figure 18-2, the sharp F1 spike is absent from the vGRF when running with a forefoot-strike pattern. In addition, the forefoot strike allows the strong muscles on the back of the ankle joint (the gastrocnemius and soleus) to contribute more to energy absorption. It has also been suggested that the forefoot running style is more efficient, because the energy stored in the strong tendon on the back of the ankle (the Achilles tendon) can be released to aid in the push-off phase of running. (I explain stored elastic energy in tendons in Chapter 15.)

Protecting Our Beans: Helmet Design

This is not a description of the biomechanics of the bean plant (although the biomechanics of fruits and vegetables is an ongoing area of active research). Instead, this section describes some of the research involved in designing helmets to protect our heads from concussions.

The brain is suspended within the skull by tough connective tissue and surrounded by a thin layer of fluid (cerebrospinal fluid, to be precise). As the head moves, the brain moves along with it. When the skull slows down, the connective tissue and fluid act to slow down the brain.

A concussion is a biomechanical event. Mechanically, an external force applied to the moving head causes it to slow down rapidly (a high acceleration), but the connective tissue and fluid aren't able to slow down the brain within the skull. The moving brain contacts the inner surface of the skull. This force, between the skull and the brain, essentially bruises the brain, disrupting the connections between the neurons.

The use of helmets in sports such as football, hockey, motorcycling, and bicycling, as well as in the military, is intended to prevent concussions. Designing better helmets is a focus of research involving several approaches:

- ✔ **Investigating the relationship between head motion, the size and direction of the externally applied force, and the force applied to the brain:** This basic research often uses cadavers, but this approach is expensive, and not enough people donate their bodies for scientific research. Researchers continue to create and refine mathematical models of the brain within the skull, allowing for computer simulation of different motions and forces.

- ✔ **Determining the relationship between the force applied to the brain and the amount of brain damage:** Some concussions are caused by an obvious, single incident like a violent collision between two players, or a player's head striking the playing surface. Other concussions are thought to occur from multiple applications of a lower level of force. It's also thought that the brain is more susceptible to a concussion from force applied in some directions.

- ✔ **Designing helmets to reduce brain loading:** Helmets have evolved from a simple protective layer of leather designed to prevent cuts and abrasions to a system intended to reduce brain loading and prevent concussions. Researchers continue to refine helmet design to

   - Spread the imposed load over a larger area, reducing the force applied to any one area of the head.

   - Redistribute the load away from areas of the head more susceptible to injury.

   - Increase cushioning within the helmet to increase the time over which the head is brought to rest, serving to reduce the magnitude of the peak and average force.

One method of testing helmets' effectiveness involves using a swinging pendulum. Researchers attach a helmet to the pendulum, pull it back, and allow it to swing into a force platform at a speed mimicking a collision typical of the sport of interest. A *head form,* designed to mimic the mass characteristics of a human head, is instrumented with sensors measuring the magnitude of force applied to different areas of the skull. The information from this testing is used to refine the design and padding of the helmet and reduce the loading imposed on areas of the skull over brain areas susceptible to concussions.

Woodpeckers have a natural helmet

Woodpeckers repeatedly slam their beaks into trees while searching for insects. When they find tasty bugs, the woodpeckers' long tongues grab it to pull it in as a meal. Woodpeckers are generally free from concussions. Biomechanists have studied the woodpecker to identify how they've adapted to avoid concussions. While the beak slams into the tree, the woodpecker's long tongue is pulled back into the head and wrapped around the brain to provide a cushioning layer of protection.

An offshoot of this research is another area of injury prevention research. In football, sensors have been attached to helmets used in games and practices. Information from the sensors is wirelessly transmitted to a receiver on the sidelines. If contact during a game causes a sensor to be loaded to a level known to be associated with a concussion, the medical staff can call the player to the sidelines to be assessed for concussion symptoms. This is important, because the risk of long-term brain damage is increased if a concussed brain receives additional trauma from a subsequent loading (in other words, if the player gets hit hard again).

Balancing on Two Legs: Harder Than You Think

For most of us, balance isn't something we have to think about. Maintaining upright posture for humans is actually a formidable challenge, but our neuro-muscular system is well adapted to the task. However, when maintaining balance is compromised, people have an increased risk of falls and subsequent injury.

The basic biomechanics of upright stance are shown graphically in Figure 18-3. A human standing upright is not motionless. Because the center of gravity of the human body is located at about waist level (for more on center of gravity, see Chapter 8), the upright human sways forward and backward, pivoting around an axis through the ankle joint (shown in the figure with ∘). Sway is a form of angular velocity (explained in Chapter 9) and is represented in the figure with the symbol ω and an arrow showing the direction of sway.

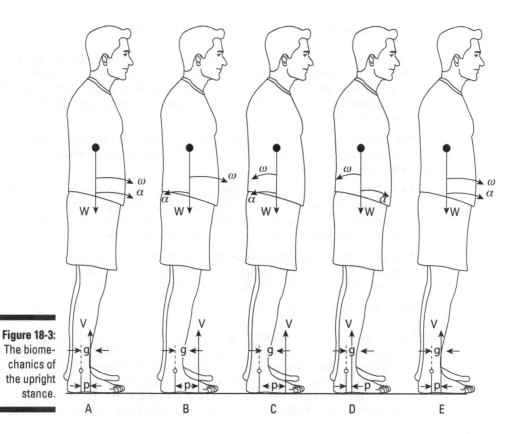

Figure 18-3: The biomechanics of the upright stance.

The external forces acting on the upright human are weight (W), applied at the center of gravity, and the vertical ground reaction force (V), applied on the feet at a point called the *center of pressure*. W is represented in Figure 18-3 with a vector (arrow) pointing downward, and V is represented with an arrow pointing upward. In upright stance, W and V are essentially equal in magnitude but opposite in direction. Each force can create a *torque,* or turning effect, around the ankle joint (see Chapter 8 for more on torque).

The magnitude of the torque created by each force is the product of the force and the *moment arm* (*MA*), or perpendicular distance from the line of action of the force to the ankle joint. In Figure 18-3, the *MA* for W is identified as g, and the *MA* for V is identified as p. The farther from the axis of rotation, the larger the torque created.

The net torque on the person is the difference between the forward torque created by Wg and the backward torque created by Vp. Because W and V are equal in magnitude, the relative size of g and p determine the net torque. The net torque causes an angular acceleration of the person, represented in the figure with the symbol α and an arrow showing the direction of the acceleration.

Here's an explanation of parts A through E of Figure 18-3:

- ✔ **A:** At this starting point, the person is swaying forward. The center of gravity is ahead of the center of pressure, causing a forward α of the forward-swaying person. To stop the forward sway, the person activates the muscles on the back of the ankle to push the center of pressure forward along the foot.

- ✔ **B:** The activated muscles have caused the center of pressure to move ahead of the center of gravity, causing a backward α and slowing the forward ω.

- ✔ **C:** The backward α has caused the person to stop swaying forward and to now be swaying backward. When backward sway is detected, the activity in the muscles on the back of the ankle is decreased and the center of pressure begins to move backward along the foot.

- ✔ **D:** The center of gravity is again in front of the center of pressure, creating a forward α of the backward-swaying person. This stops the backward sway and reverses it to a forward sway.

- ✔ **E:** This is the same as part A.

Balance is a repetition of the sequence A through E. The sensory receptors of the central nervous system must be able to detect sway and coordinate the activation and deactivation of the muscles to control the angular acceleration. The muscles must be able to produce the required pull on the foot to move the center of pressure forward to stop forward sway. Any disease or injury of the neuromusculoskeletal system can affect the control of balance.

Using a force platform, the location of the center of pressure can be identified and recorded. The tracing of the movement of the center of pressure under the foot is called a *stabilogram,* and analyzing the stabilogram shows how an individual controls balance.

One use of the stabilogram has been to evaluate the effectiveness of spongy floor mats used by people who work standing up at workstations for long periods of time, like cashiers, assembly-line workers, and kitchen staff. These workers tend to develop soreness in the ankles and feet because of blood pooling in the veins of the legs. Getting blood back to the heart depends on active muscles helping to pump the blood back up the veins against gravity (for more on circulation, see *Anatomy & Physiology For Dummies,* 2nd Edition, by Maggie Norris and Donna Rae Siegfried [Wiley]). The stabilogram confirmed that using a cushioned floor mat tends to slightly destabilize the upright worker, causing greater sway. Combining electromyography (EMG; see Chapter 17) with the stabilogram measures confirmed that greater muscle activity in leg muscles accompanies greater sway. It's important that the amount of cushioning be just enough to increase sway to require more muscle activity, but not so destabilizing that the muscle activity leads to earlier fatigue.

Chapter 19

Investigating Forensic Biomechanics: How Did It Happen?

. . . .

In This Chapter

▶ Gathering the information you need

▶ Determining how the person was injured

▶ Evaluating conflicting versions of what happened

. . . .

*F*orensic biomechanics uses the principles of biomechanics to answer questions raised in criminal and civil law. The principles of biomechanics are those describing the kinematics and kinetics of moving bodies and tissue responses to loading. (See Parts II and III for more on linear and angular kinematics and kinetics, and see Part IV for the biomechanics of materials.)

A forensic analysis is typically used to answer a question related to how somebody was injured or killed. The answer has legal implications. In a criminal case, such as an investigation of an assault or murder, the analysis is used to describe the scenario involved with the alleged criminal act and to explain how it happened. The question of whether an accused individual was capable of physically performing the act is usually also addressed. In a civil case, the purpose is to describe the cause of injuries incurred during an accident. A dispute has to be resolved regarding how the injuries were sustained, and the analysis is used to decide if a flaw in design or procedures makes the defendant (the accused) negligent and liable for financial compensation to the plaintiff (the accuser).

Forensic biomechanics isn't part of the diagnosis, treatment, or medical management of injuries — those are the responsibilities of the medical community. Forensic biomechanics doesn't decide liability and doesn't get into issues of whether an action was legal — those are the responsibilities of the legal process. Instead, the outcome of a forensic analysis is a professional opinion on what occurred, an opinion supported by the principles of biomechanics.

In this chapter, I briefly describe the steps involved in a forensic analysis and provide examples of applying the analysis. This chapter is just a brief overview of a growing field — entire books have been written on the subject. If you're interested in learning more, check out *Forensic Biomechanics,* by Jules Kieser, Michael Taylor, and Debra Carr (Wiley).

Collecting Information for a Forensic Biomechanics Analysis

Rarely is a forensic biomechanics investigation begun until considerable time has passed since the incident occurred. Because conditions at the site of the incident may have changed, the analysis relies on records compiled in an investigation of the incident, pulling the information together under the principles of biomechanics. This is often referred to as a *reconstruction of the event.* It leads to the development of a biomechanically sound scenario that, in the analyst's professional opinion, explains what happened. It can also include an opinion as to why an alternative scenario developed by another analyst or provided by the defendant or the plaintiff did not occur.

The records used in a reconstruction are the same records underlying conflicting opinions. I cover what's typically included in those records in the following sections.

Witness accounts

Witnesses to the incident are interviewed by police, and their personal opinions of what happened are recorded. Witness accounts from the general public are generally the *least* reliable of the documents reviewed because a personal opinion may be tempered by the person's attachment to the incident. Statements could be worded so as to avoid suspicion and incrimination or to make the individual appear more heroic. Witness accounts can also be affected by the shock of the incident.

First responders (including the emergency personnel called to the scene, typically both rescue and medical teams) provide details on the environmental conditions. Police take their statements as part of an investigative report.

It's important that the first-responder statements be as thorough as possible, because first responders are often asked to provide a statement under oath if a civil or criminal investigation proceeds. Between the time of the incident and the deposition or trial, a lot of time may pass, and the first responders will serve at multiple other accidents in between. This may make it difficult for them to recall details that weren't written down as soon as possible after the incident occurred.

Initial statements from the victim and witnesses can affect how the scene is treated by first responders. If a fall victim says, "I was pushed," first responders will treat the scene as a possible crime scene and avoid damaging any potential evidence during treatment and removal of the victim.

Police incident investigation reports

Police responding to the scene typically complete an *incident investigation report,* which includes the officers' initial perception of the scene, statements from witnesses, and documentation of the scene (including measurements and possibly photographs). If criminal activity is suspected, the police forensic unit measures and documents the scene more extensively, including recovering clothing traces and taking swabs of body fluids and scuff marks for DNA testing.

The police create a detailed sketch with measurements showing the physical features of the scene, the location of the victim(s) when first found, and the position of potential evidence related to the incident. Thorough documentation is important because the scene may change following the incident, including removal of objects, cleanup of debris, and alterations because of seasonal change, such as plant growth.

Medical records

Medical personnel treating the injured individual, often at the emergency room, compile a list of injuries. Medical personnel focus on the diagnosis and treatment of conditions, but they have limited training in what caused the injury in the first place. Medical records rarely address the biomechanics of how the injury was sustained, other than a blanket statement such as "Patient was involved in a motor vehicle accident," "Patient sustained injuries during an assault," or "Patient suffered an injury related to his work."

The medical records include the physician's written documentation of the injuries, including X-rays and other internal imaging used in the diagnosis.

The listed injuries include the location and description of bone fractures, joint dislocations, cuts and bruises on the skin, and trauma to internal organs.

If a criminal act is suspected, a police photographer may record the external injuries. These photographs are important because medical records may document and focus only on the more serious injuries. What a doctor may consider minor abrasions and contusions won't necessarily be part of the medical records, but they could be important evidence for forensic analysis. It's also important to keep the clothing of the patient, or at least photograph the clothing, so that marks useful in a forensic analysis are available.

The medical records list height and weight, as well as identify pre-existing conditions (including degenerative conditions such as low bone mineral density, low muscle mass, obesity, and previous surgery). All these factors affect the individual's ability to withstand traumatic loading.

If the patient is dead or dies, an autopsy will be performed. Not all autopsies are viewed as equal in the legal community. Some are supervised or conducted by an elected coroner, who may have minimal medical training. The most accepted autopsy is one conducted by a *forensic pathologist,* a physician specially trained in determining cause of death. A forensic pathologist is sometimes called a *medical examiner.* The autopsy report contains an extensive listing of the external and internal injuries and pre-existing conditions and specifies the cause of death.

Determining the Mechanism of Injury

The medical records identify the injuries sustained. A forensic biomechanics analysis works backward to identify and describe *how* the injuries were sustained, called the *mechanism of injury.* There are basically three main steps in this process:

1. **Identify the type of loading known to cause the recorded injuries.**

   The types of loading include tension, compression, bending, shear, and torsion. See Chapter 12 for more on the types of loading (stress) and the tissue response (strain).

   The medical records list the type and location of injuries sustained, and the injuries must be matched to how a tissue or anatomic structure is known to fail. The mechanics of materials describes the stress–strain relationships of different tissues and anatomic structures like bones, joints, ligaments, skin, blood vessels, and internal organs. The description links the recorded injuries and the type of loading that had to be present.

Specific types of loading applied to specific tissues or structures cause specific injuries. For example, high levels of *compression* (squeezing load) to the rib cage cause rib fractures and *contusions* (bruising) and *lacerations* (cuts) to internal organs, including the lungs, heart, liver, and spleen. Severe dislocations and fractures close to the ankle joint occur from high *tension* (pulling load). Fractures of the *vertebrae,* the bones of the spinal column, occur in different parts of the bone depending on whether the loading was compression or *torsion* (twisting load). A *shear* (sliding or rubbing) stress on the skin causes a different abrasion than does a compression load. The description of the injuries in the medical records is used to identify the type of loading that must have occurred during the incident being reconstructed.

The published literature — including textbooks, technical reports, and scientific journals — provides the information relating types of loading to types of injury. It's not necessary to memorize each injury-loading relationship, but it is necessary to know how and where to find reputable and accepted sources of the specific information.

2. **Identify the other body that potentially interacted to produce the injury-causing load.**

   Two bodies must interact in order for a force, and loading, to occur; one of those bodies is obviously the injured part of the victim. The details provided in the description of the incident site are critical to identify potential bodies that could impose the type of loading required to produce each recorded injury.

3. **Describe the specific event causing the load responsible for each injury.**

   In other words, which one of the potential bodies caused the injury (sometimes called the *involved physical component*), and how did the motion of the two bodies lead to the contact producing the load? This is the actual reconstruction of the events, producing a scenario explaining sequentially all the injuries sustained. The scenario must account for each injury.

In the case of motor vehicle collisions, an extensive database compiled by transportation safety groups helps with this step. Decades of vehicle crash testing using *anthropometric models* of the human body (so-called "crash-test dummies") and studying the injuries sustained in actual single and multiple vehicle collisions have revealed patterns of injury and loading during different types of collisions, including head-on, *T-bone* (where one vehicle hits the other from the side), and rollover. The database provides a good starting point for reconstructing a scenario by outlining the expected types and patterns of injury during motor vehicle collisions.

The developed scenario can't violate the principles of mechanics, starting with Newton's three laws of motion (explained in Chapter 6):

- ✔ The motion of a body remains constant unless acted on by an unbalanced external force (Newton's first law).

- ✔ An unbalanced external force causes an acceleration (speeding up or slowing down) of the body (Newton's second law).

- ✔ The force on each body is equal in magnitude but opposite in direction (Newton's third law).

Identifying the specific object involved in causing a specific injury is facilitated by Newton's third law. Because a force is applied to both bodies, both bodies will be stressed. The documentation of the scene often shows objects or structures that are deformed. Matching the deformation to specific body parts injured can identify the involved physical component.

Newton's laws of motion are the basis of all motion, and specific kinematic and kinetic principles explained in Parts III and IV of this book are used to show the movement of the body or bodies leading to the injuries. For example:

- ✔ A body in free fall — whether it's a driver ejected from a vehicle, a worker falling from a ladder, or a victim thrown from a height — follows the parabolic motion of a projectile (see Chapter 5 for more on projectile motion). The shape of the parabola is determined by the horizontal and vertical components of the resultant velocity and by the height of the body at the instant it became a projectile. The equations of projectile motion are combined with the involved physical components encountered along the trajectory to identify the kinematics of release (velocity and position).

- ✔ Kinetic energy is increased when a body moves faster, and work must be done on the body to change its kinetic energy. The work–energy relationship (explained in Chapter 7) is of particular importance when looking at how a moving body is brought to rest. If the work is done over a short distance, such as a faller landing on the flat concrete surface of a basement floor, the force magnitude is higher than if the work is done over a longer distance, such as the faller first striking the third stair from the bottom and rolling down to the basement floor. The type and pattern of injury to different body structures produced by low-force/long-distance changes in kinetic energy differ from those produced by high-force/short-distance changes in kinetic energy. It's not how fast the body moves that causes injury; it's the way the work is applied to bring the body to rest.

✔ If an applied external force creates an unbalanced torque on a body, the body undergoes *angular acceleration* (speeds up or slows down the rate of rotation, as explained in Chapter 10). This applies to the human body as a whole, as it rotates around its center of gravity while a projectile, and to individual segments rotating at joints. Points farther from the axis of rotation have a greater linear velocity than points closer to the axis of rotation, as explained in Chapter 9. The rate of rotation and the point of contact affect the magnitude of the force imposed when a body or segment strikes another body.

Combining the principles of biomechanics with the medical records produces a scenario explaining the motion of the body and the cause of each injury. The analyst's level of confidence in each step of the scenario is expressed as "possible," "probable," or "certain." The legal process then considers the scenario, along with other evidence, to come to a legal decision.

Evaluating Different Scenarios

Forensic biomechanics addresses the question of "What happened to cause the injury?" The incident has already occurred, and opinions differ on the injury-causing scenario. Both the plaintiff and the accused will have an opinion. And you can bet that both opinions are tempered (whether intentional or not) by at least a little self-serving bias. The reconstruction of the incident in a forensic biomechanics analysis is intended to provide an objective description of the event to a level of confidence based on the principles of mechanics and available evidence.

Conflicting interpretations of the evidence can lead to different scenarios describing the same event. In this section, I walk you through two examples where forensic biomechanics is used to come to resolution.

Ending up on the far side of the road

A semi-truck going north on a foggy two-lane asphalt highway hits a passenger car at a T-intersection. The car is crossing the northbound lane to turn south on the highway. The collision causes the car to travel north while spinning off to the shoulder of the road, and the semi veers off the road and overturns. The first responders find the driver of the car, an obese man, lying on the gravel shoulder of the highway across from his vehicle; the driver gives two gasps of air, but no further response, and is pronounced dead. Inspection of the vehicle reveals extensive damage inside the passenger compartment and no indication of seat-belt use.

The police report claims that the driver was ejected from the vehicle — a reasonable conclusion based on the differing locations of the driver and the vehicle, and the lack of seat-belt use. A reconstruction of the collision by automotive engineers indicates two points during the spin of the driver's car when he could have been ejected.

Using the data from the automotive engineers, the driver's horizontal and vertical velocity and height of release were calculated at the two potential points of ejection. The projectile motion of the driver calculated at either point showed that he would land on the asphalt, and then bounce, slide, and/ or roll to the gravel shoulder. These calculations supported the claim that the driver could have been ejected.

However, the medical reports and photographs showed no significant *skin abrasions* (deep skin injuries caused by friction, including embedded road-way materials, often known as "road rash") or shredded clothing, both of which would be expected if the driver had been ejected, landed, and bounced/rolled/slid on the asphalt and gravel. The injuries listed in the medi-cal records — including contusions, fractures, dislocations, and damage to the heart and liver — were all explained by the unbelted driver remaining in the vehicle between the time it collided with the semi and the time it came to rest. The interior damage to the vehicle matched the injuries. The lack of injuries consistent with ejection led to a decision that the driver exited the vehicle after it came to rest, and collapsed on the shoulder after walking across the highway.

Automotive engineers reconstruct the motion of vehicles before, during, and after a collision using the same principles of mechanics as used in a forensic biomechanics analysis. However, automotive engineers use the material prop-erties of the vehicles and components rather than the material properties of biological materials in their reconstruction.

Landing in water with a broken jaw

The body of a 1.5-m-tall (5'), 500-N (110-pound) female was found face down in a 0.25-m (10-inch) pool of still water 4.6 m (15 feet) below the railing of a walking bridge over a gully. The bridge railing rises 0.94 m (37 inches) above the bridge walkway. A steel support bar from an earlier bridge runs across the gully parallel to the walking bridge, 1.5 m (5 feet) out and 2.4 m (8 feet) below.

On questioning by police, her 1.8-m-tall (5'11") male companion first claimed that she was dancing on the bridge and fell over the railing. However, fol-lowing reports that they had been seen fighting in a nearby residence, he claimed that they argued on the bridge, he pushed her backward, and she flipped over the railing. The location of her body supported either scenario, but the different versions and the circumstances prompted a forensic biomechanics analysis.

The female's medical records listed a deep cut on the bottom of her chin, a broken jaw, an angled line of bruising over four broken ribs on her left side, and bruising under the left armpit. Falling and landing in the water would not cause these injuries. The water would increase the distance she was brought to rest, and the force of the water bringing her falling body to rest would not break her jaw or ribs. The fractures and bruising indicated she had struck the steel support bar with her chin and left side. She had to have gone over the railing with enough horizontal velocity to travel 1.5 m (5 feet) out to strike the bar while falling the 2.4 m (8 feet) down.

If she had been dancing or pushed backward, the victim's height would not allow her to tumble over the railing. Her center of mass (explained in Chapter 8) would be below the top of the railing, which would keep her on the walkway. Using the equations of projectile motion along with the height and build of the male, a scenario was developed showing that he lifted her up to hold her horizontally across both forearms, stepped forward, and released her over the railing. When prosecutors described the scenario to the male, he admitted to the scenario, saying he had done it because he wanted to "teach her a lesson" but said that he was unaware of the steel support bar.

If you're considering a career in forensic biomechanics

Forensic biomechanics is a specialized area of practice, applying the principles of biomechanics in legal situations. If you want to pursue a career in this exciting field, you'll need a strong academic background, typically a minimum of a master's degree focused on biomechanics, with coursework (and practical experience) in kinematics, kinetics, and mechanics of materials (especially the components of the neuromusculoskeletal system). You'll also need strong analytic and interpersonal skills, because you'll typically work with a group of other scientists, lawyers, engineers, and technical staff (including animators) to produce a scenario, with each partner relying on the expertise of the other members of the group to re-create the incident. A thick skin, calm demeanor, and ability to think on your feet are necessary, because in a legal case, everything gets challenged, from your credentials to analyze the evidence to the version of the incident you produce.

Part VI

The Part of Tens

For the ten basic principles of biomechanics, head to www.dummies.com/extras/biomechanics.

In this part...

- ✔ Find informative websites on biomechanics.
- ✔ Discover some things you may not know about biomechanics.
- ✔ Get tips for succeeding in your biomechanics course.

Chapter 20

Ten Online Resources for Biomechanics

*T*his book is a fantastic resource on biomechanics (if I do say so myself), but don't stop reading about biomechanics when you finish reading this book! A treasure-trove of information is available online. In this chapter, I introduce you to some of my favorite sites.

The Exploratorium

www.exploratorium.com/explore/staff_picks/sports_science

The Exploratorium is a "hands-on" science museum in San Francisco, a great place to visit if you happen to live near or be traveling to the City by the Bay. The Sports Science portion of the Exploratorium's website includes interactive pages explaining and demonstrating the biomechanics of different activities, including cycling, skateboarding, baseball, and hockey. Be sure to explore the many other links on this site — they're all fascinating, including several on the human body.

The Physics Classroom

www.physicsclassroom.com

This website is very popular among the students in my biomechanics courses. It was developed as a resource for teaching physics, including mechanics, at the introductory level. The site includes several features that make it a valuable tool for learning about *kinematics* (position, velocity, displacement, and acceleration); force, Newton's laws; and work, energy, and power.

The Physics Tutorial link takes you to individual lessons that explain the basic concepts and provide really helpful graphics. The "Check Your Understanding" section at the end of each lesson includes review questions with step-by-step solutions.

The Minds On Physics link takes you to a set of multiple-choice questions related to the different concepts. You get instant right/wrong feedback on your response, and if you get one question wrong, a similar question pops up later. If you miss too many questions on a single topic, you'll be referred to a topic-specific Help section.

The Calculator Pad link takes you to 25 to 35 word problems on each of a variety of concepts. You can check your answers and, if you want, listen to an audio file describing how to solve the problem.

Finally, click the Review Session link to be taken to review questions on the different concepts. You can select the concept you want to review. The review questions are provided on one page, and the questions are linked to the answers, explanations, and solutions.

Coaches Info

www.coachesinfo.com

This website provides information and education for athletes, coaches, parents, and teachers. The available instructional materials typically include a focus on the biomechanics of performing the activity. Two of my favorite links are

- **Sports to Choose From:** Hover your mouse over this menu item, and you see a variety of sports, including swimming, diving, golf, gymnastics, tennis, track and field, basketball, and triathlon. Within each sport are additional links to short, easy-to-read "how to" articles written by an expert in the activity. Most of these articles include some aspect of biomechanics, whether it's leg motion during the punt in rugby, forces during the vault in gymnastics, or arm action during the volleyball jump serve.

- **News from the Knowledge Factories:** Click this menu item to find links to news from different sports research centers (the "knowledge factories" referred to in the title). The available material is frequently updated, and the "news" may be a recording of a presentation given by a researcher discussing the biomechanics of a specific sport or activity.

Textbook-Related Websites

When you purchase a biomechanics textbook, you may be assigned (or be able to purchase for a small additional fee) a code giving you access to a website developed as a supplement to the textbook. The site is maintained by the publisher. Follow the instructions provided with the textbook to access the site.

Most textbook sites typically include solutions to all the questions at the back of each chapter (with some showing the step-by-step guide to the correct answer), additional quizzes, and links to online resources providing more information on topics raised in the textbook. Use this companion website to your advantage!

Also available on some textbook sites is access to video recordings of different sports and activities. Recent editions of most textbooks provide access to an easy-to-use software program for conducting a biomechanics analysis, to be used with the video recordings.

Topend Sports

www.topendsports.com

This site calls itself the "The Sport + Science Resource." Select the Science tab on the home page, and you're taken to a list of different aspects of sports science, including biomechanics.

The Biomechanics menu provides a variety of choices, including Physics of Sport and, my favorite, Sport Specific. The choices available on the Sport Specific page provide interesting facts related to different sports, as well as an overview of some basic data collection techniques (including how to calculate joint angles from video on your TV using pens and plastic sheets).

Dr. Mike Marshall's Pitching Coach Services

www.drmikemarshall.com

Mike Marshall had an outstanding career as a relief pitcher in the major leagues of baseball, winning the Cy Young Award as the best pitcher in 1974.

While still playing in the major leagues, Dr. Marshall pursued studies in exercise science and received his doctorate in biomechanics in 1978. A star pitcher with an advanced degree is quite a unique combination.

Dr. Marshall's career was frequently interrupted by arm injuries. Using high-speed cameras, he has recorded and analyzed his own throwing motion and that of other pitchers, and you can watch these videos in slow-motion. During the pitching motion, the arm goes into extreme positions that create high stress in muscles, ligaments, tendons, and bones. Because pitchers repeat the motion frequently and at high intensity, all these structures are at risk of injury.

The site also explains how Dr. Marshall used his knowledge of biomechanics to develop an alternative style of pitching, which is explained on this site. Marshall's technique is intended to allow a pitcher to perform as well as the traditional pitching motion without imposing the same stress on the body. You can compare the high-speed recordings of the traditional and alternative pitching techniques.

Although no pitcher using his method has reached the major leagues (blamed in part, by Dr. Marshall, on an unwillingness of traditionalists to accept his different technique), the site is at least a very thought-provoking example of how biomechanics can be used to suggest a safer ways to perform at a high level in sport.

Waterloo's Dr. Spine, Stuart McGill

`http://www.youtube.com/watch?v=033ogPH6NNE`

Low-back pain is a common problem among both active and sedentary individuals. "Safe" techniques for lifting and exercises for back health have been recommended for a long time, but the incidence of low-back problems remains high among people of all age groups.

For 30 years, Dr. Stuart McGill and his colleagues have investigated the biomechanics of the back. Their research has further explained the neural, muscular, and skeletal anatomy of the fascinating and complex human back. In addition, Dr. McGill has analyzed the biomechanics of most exercises and lifting techniques recommended for back health. The studies include measuring muscle activity and recruitment patterns using EMG, and calculating the stress at frequently injured structures of the back. The research has challenged many of the traditional ideas about low-back health, exercise, and safe lifting technique.

Dr. McGill was one of the first to recommend that the objective of low-back exercise should be to improve the stability, rather than the flexibility, of the back (now commonly referred to as *core stability*). In this four-minute interview, Dr. McGill points out what's biomechanically wrong with some traditional back exercises (including the standard crunch and sit-up, which may actually cause low-back pain). More important, he suggests alternatives that are more biomechanically sound. The video includes brief demonstrations of the alternative exercises, along with an explanation of why they're better.

Skeletal Bio Lab

`www.fas.harvard.edu/~skeleton/`

Dr. Daniel E. Lieberman is the head of the Skeletal Bio Lab, which studies the evolution and function of the musculoskeletal system. The lab is part of the Department of Human Evolutionary Biology at Harvard University, a group of researchers and educators interested in the question "Why are humans the way we are?"

Among the multiple links available from the main page of the Skeletal Bio Lab, my favorite is Barefoot Running (`www.barefootrunning.fas.harvard.edu`). This site provides a great overview of the biomechanics of, as the name of the link implies, running barefoot. The site explains, and includes great slow-motion videos showing, how running with shoes alters the biomechanics of the foot and lower limb compared to running without shoes. Humans have been running for millions of years, far longer than running shoes have been available, and Lieberman's group suggests that early humans ran, without shoes, differently from modern humans.

Biomch-L

`http://biomch-l.isbweb.org/forum.php`

This site is a discussion group sponsored by the International Society of Biomechanics. Participants include clinicians, researchers, instructors, and students. Postings come from individuals asking questions, seeking advice, or raising a topic for discussion. Some postings list employment opportunities or graduate study opportunities, indicating the required qualifications. This provides a resource for students wondering how to prepare for a future career in biomechanics.

Before you post on this site, do a little background searching first. Refer to information you've found related to the question, and then explain why it's incomplete and leads to your posting the question. Don't simply post a problem from class and ask for help — Biomch-L is not a worldwide access to "office hours" from experts!

One of my favorite discussion topics on Biomch-L (available from the archives) was focused on the fairness of allowing individuals with prosthetic limbs to participate in competitions against non-amputees. The discussion raised issues related to biomechanics, physiology, and ethics of allowing this mixed competition.

American Society of Biomechanics

www.asbweb.org

The American Society of Biomechanics (ASB) is a professional group founded in 1977 to promote the growth of research and education in the field of biomechanics. The society includes biomechanists working in a lot of different areas, including exercise and sport science, biology, orthopedics, and engineering.

The ASB holds an annual meeting, providing an opportunity for biomechanists from the different areas to come together and share ideas. If you click the Conference Archives link, you can access hundreds of different research projects that have been presented at the annual meetings since 1999. These are available as two-page *abstracts,* or overviews, of the projects. The archives are a simple way to see how biomechanics is used to answer questions from multiple areas, from sport to aging to evolution.

One of the most fascinating (and most popular) links from the ASB site is the Video of the Month (you can access this directly at https://sites.google.com/site/biomechvideo/). Although the site isn't always updated each month, the collection of videos includes cats, humans, cockroaches, geckos, and robots demonstrating principles of biomechanics through a wide variety of activities. The videos are entertaining, even to a non-biomechanist.

Another valuable resource is the Students page (www.asbweb.org/html/students/student.html), which includes additional links to graduate programs in biomechanics and some links to classes on biomechanics.

Most of the site is open to non-members, but you can apply for membership in the ASB, too. If you live outside the United States, you can find links to biomechanics societies in other countries and regions. The ASB site also links to other professional groups like the International Society of Biomechanics and the International Society for Biomechanics in Sport.

Chapter 21

Ten Things You May Not Know about Biomechanics

In This Chapter

▶ Appreciating the reach of biomechanics

▶ Discovering facts to impress your friends

T he principles of biomechanics apply to everything that moves. In this chapter, I fill you in on ten interesting things about biomechanics that you may not have realized. Enjoy!

Looking at How Biomechanics Got Its Start

Giovanni Alphonse Borelli (1608–1679) is known as the "father of biomechanics." Borelli was a physiologist and physicist most famous for two volumes: *On the Motion of Animals I* and *On the Motion of Animals II*. In these volumes, Borelli applied principles of physics, including line of action of force and torque, to describe the musculoskeletal system of animals. Borelli was the first to accurately explain that muscle can produce only a pulling force, and he first described how the lower extremities provide support for the body while walking.

Borelli's sketches in the volumes are essentially free-body diagrams of the skeletal system. The diagrams show how the fixed attachment points of a muscle to the bony segments create a moment arm between the muscle force and the axis of rotation of the joints the muscle crosses — the idea that muscle force serves as a torque generator at the joints (see Chapter 8). Although he was long dead by the time the American Society of Biomechanics was founded in 1977, the Society's highest award is called the Borelli Award in his honor. The Borelli Award is given to a biomechanist for an outstanding career of contribution to knowledge of biomechanics.

Adding Realism to Entertainment

Videogames and movies use computer animation to add action and adventure to the gaming and movie-going experience. Movie studios and videogame producers don't want to expose live actors to dangerous situations that are written in the script (well, maybe they *want* to, but they'd go through actors like Kleenex, and who wants to mess with that?), so they can make the escapades of the characters much more spellbinding with animation.

The data collection hardware and processing software used for movies and videogames is similar to that used in a biomechanics lab for research (see Chapter 18). In a research project, data are collected to answer a specific question. In the entertainment industry, data are collected from athletes and actors for use in the animation.

In a motion capture studio, an athlete or actor is recorded performing a variety of activities. The recording is *digitized* (processed into a format that can be manipulated with computer software, as explained in Chapter 17). Then animators can use computer software to manipulate the digitized recording to re-create a physical image of the performer performing different activities from those originally recorded — some too dangerous for a live actor. In a videogame, the sophisticated gaming software allows you, from the comfort of your couch, to manipulate the characters as part of the videogame storyline.

In 2005, a group of biomechanists and engineers received a Technical Achievement Award, an Oscar, from the Academy of Motion Picture Arts and Sciences for their contributions to improving the quality of the hardware and software used for computer animation.

Developing Safer Motor Vehicles

Occupants of a motor vehicle are much safer now when things go wrong than they were in the past. Safety features built into the vehicle frame and exterior are designed to absorb energy during a crash and maintain the integrity of the interior occupant compartment. In addition, the interior compartment of most vehicles includes safety features such as air bags; a padded dashboard; breakaway steering columns; and shatter-resistant, energy-absorbing windows. All these features are intended to reduce the forces imposed on an occupant during a crash by increasing the time to bring the occupant, or a body segment, to rest and by spreading the imposed force over a larger area of the body. Biomechanics research has been an important part of designing these safety features, and the research typically includes the following four parts:

✔ **Identifying the mechanism of injury, or how an occupant gets injured:** Research involving controlled crashes in laboratories and collision reports are used to identify how occupants get injured in a crash. When the force of a collision changes the motion of a vehicle, the occupants continue to move until an unbalanced force acts on them (the law of inertia; see Chapter 4). Reconstructing the accident shows how the occupant moved after the collision and identifies what part of the vehicle interior (seat belt, air bag, door panel, dashboard, windshield, and so on) applied force to different segments of the occupant (arm, leg, trunk, head, and so on), and how much force was applied. *Crash-test dummies* (sophisticated models of the human body including sensors to measure loading of different body parts) are used in lab research to obtain this information.

✔ **Identifying the response and failure levels of different body segments, tissues, and organs:** This information comes from testing donated cadavers in a lab, and it's needed to identify how much stress different body parts can withstand. The objective of a vehicle safety feature is to keep the level of loading during a crash below the level that will cause serious injury, especially to critical body parts like the brain, heart, and other internal organs. So, it's important to know the levels of stress that must be avoided.

✔ **Designing safety features to provide protection:** With information on what parts of the vehicle interior apply load to a body segment, and how much load a body segment can withstand before sustaining a serious injury, the vehicle interior and the occupant restraint system can be improved. Such safety features as shoulder straps on seat belts (even in the back seats), air bags in the side panels of doors and windshield supports, and better child seats have been developed based on this research.

✔ **Developing computer models of vehicles and occupants:** After the information is obtained on how people move within a vehicle during a crash and how tissues are loaded, a computer model of this information can be developed. Crashing cars and building crash-test dummies is expensive. With a computer model, researchers can more easily and less expensively simulate a crash and use the results to improve occupant safety.

Improving the On-Shelf Quality of Fruits and Vegetables

The *bio* in biomechanics refers to life. Fruits and vegetables are living things (although most of us can't have a conversation with one), and the laws of biomechanics are used to improve the quality of the products for sale in stores.

Most fruits and vegetables are grown in one place and transported to another for sale. Shoppers prefer good-looking produce, free of blemishes and soft spots. If the produce is too roughed up, people pass it by, creating a loss for the store.

The handling required during picking, processing, shipping, and displaying can damage the fruits and vegetables. Just like tissue in the human body, the damage on a fruit or vegetable comes from the stress and strain it undergoes from field to store. To reduce damage, the tolerances of fruits and vegetables are determined, and better methods are developed to reduce the load imposed between the field and the store. In some cases, the tolerance of a fruit or vegetable can be modified by selective cross-breeding, further reducing the risk of damage.

Fitting Footwear to the Activity

Many pairs of shoes are purchased simply because the buyers find them comfortable, like how they look, and can afford them (or at least not suffer sticker shock). All these considerations are worthwhile. But choosing shoes should *also* consider the fact that the shoe is the interface between the user and the surface, and as such, it can affect performance and the risk of injury.

Biomechanics plays a large role in designing specialized footwear to meet the demands of different activities and sports. Research in many biomechanics labs measures the required motions of the foot during an activity; the friction demands of starting, stopping, turning, and pivoting on different surfaces; and the forces from the ground during any jumping and landing in the activity. Data are collected from a variety of people performing the activity using motion and force-recording systems (see Chapter 17). The data are used by shoe companies to design shoes that facilitate the movements and protect the feet.

Some examples of biomechanics-based design features for sports footwear include flexible shoe soles allowing the forefoot and rearfoot to move naturally when running, enhanced shock-absorbing material under the ball of the foot for basketball shoes, and a different arrangement and sizing of cleats on shoes used for baseball and softball.

Biomechanics research has also guided the development of shoes for people with diabetes. Diabetes affects both the sensory system and the circulation system. A high-pressure area on the foot can lead to the development of an open sore that can easily become infected and risk, literally, both life and limb. Shoes designed to eliminate the high pressure from an at-risk area of the foot while still allowing mobility are prescribed to reduce this risk.

Banning Biomechanically Improved Sport Techniques

Sport biomechanists are interested in improving the techniques used to perform different events. In some cases, the analysis has led to recommendations for technique changes so different that the sport governing body changed the rules of the sport to disallow use of the improved technique. Here are three such examples:

✔ In swimming events, the drag produced by a swimmer moving through the water is a resistive force to be overcome (see Chapter 11 for more on drag). The drag force is less when the swimmer is completely submerged than when the swimmer is on the surface. To reduce the drag effect, it was recommended that a swimmer stay underwater and kick for a longer time after diving in at the start of the race and after performing a turn at the end of the pool. Elite swimmers were able to stay submerged for a long distance, actually completing most of each lap underwater. In response, the rules were changed limiting the distance and number of kicks a swimmer is allowed to perform before resurfacing after a start or a turn.

✔ The technique used for high jumping has changed over the years, and world records for height jumped have increased. One influence on technique used by a high jumper was the material in the landing pit. Originally, a shallow pit filled with loose sand or sawdust was used. For a safe landing, the jumper typically landed feet first and rolled to come to rest. After thick foam landing mats replaced sand and sawdust pits, jumpers no longer needed to land feet first for safety. The American high jumper Dick Fosbury introduced the flopping technique and won the gold medal in the 1968 Olympic Games. In the so-called "Fosbury flop," the jumper approaches the bar at an angle, and twists during the takeoff to go over the bar head and shoulders first in an arched position, facing up to the sky. The jumper lands on his or her back on the foam mat, safely brought to rest as the kinetic energy of the falling jumper is converted to strain energy of the mat (for more on energy conversion, see Chapter 7). Analysis showed the center of gravity of the jumper passed closer to the bar with this technique. Because the upper body can bend forward (flex) more than it can bend backward (extend) in the arched position, a new technique was suggested. The jumper ran straight at the bar and used a two-legged takeoff to jump up and forward, essentially performing a front somersault, or forward roll, over the bar. This technique was so radically different from the traditional form that the rules of high jumping were changed and now a one-leg takeoff is required.

✔ In the long jump, the jumper sprints toward a pit and jumps from a takeoff board to travel forward and land in the pit. At takeoff, the jumper develops a lot of angular momentum because of the torque produced during the push on the board (see Chapters 8 and 10 for more on torque and angular momentum, respectively). Various techniques are used by jumpers to stay vertical while soaring through the air during the long jump. Some jumpers adopted a somersault technique to perform the long jump. After taking off from the board, instead of trying to stay vertical, the jumper tucked and performed a forward somersault in the air before landing feet first in the pit. The angular rotation developed during the push off from the board became a positive rather than a negative effect. Because of a fear of serious injury with the somersault technique, it was banned from use in the long jump. (The fear was not from the jumpers but from observers unaccustomed to the technique. My guess is that these observers had never watched a gymnastics meet, where even more complex turning and twisting moves through the air follow a run across the mat.)

Re-Creating Dinosaurs

Dinosaur bones have fascinated people since the first scientific discovery in the early 1800s. Bones dug from an excavation site, or found after being exposed by wind and water erosion, are all that remain of these ancient animals. Often only a partial skeleton, or even only a partial bone, is recovered.

Biomechanics is used to re-create the shape of bones missing from the skeleton and to determine where the muscles were once attached to the bones. Wolff's law (see Chapter 13) states that the external shape and internal structure of a bone reflects its loading from external forces, including tensile forces from muscle (see Chapter 15). To reconstruct the muscle structure that gives the dinosaur its ultimate shape, researchers first identify the prominences on the bone serving as the attachment sites for the muscles and then determine how big the muscle would have had to be in order to create the prominence and move the animal. Differences in the size of the forelimbs and hindlimbs also reflect the loading of the bone during locomotion and support, and these differences are used to determine if the dinosaur walked upright on two limbs, on all four limbs, or in some combination of upright and four-limb gait.

Designing Universally and Ergonomically

Universal design refers to developing or redeveloping products and the built environment of everyday life to fit the physical capabilities of all people. For tools and equipment in the workplace, a similar process is called *ergonomic design.* The goal of both is to improve the utility of the product and to reduce the risk of injury when the product is used.

Essentially, the idea behind universal and ergonomic design is to make things usable by people regardless of their ability. Instead of designing the product simply to perform a task, the product is designed to match the people who will use it to perform the task. The dimensions and weight of a well-designed product can be used by as many people as possible without adaptation.

Biomechanics is an important part of the universal/ergonomic design process. The physical dimensions, or *anthropometrics,* and the strength, flexibility, and endurance limits of potential users, including those with a disability or disease, are considered in the design.

Example products of universal design include such simple tools as the potato peeler with a thicker cushioned handle, originally designed for use by individuals with flexibility and strength limitations in the hand, and more complex redesigns such as the elimination of the front steps and the installation of wider doorways in retirement community housing, intended to ease access for individuals using wheelchairs.

An example product of ergonomic design is the availability of power screwdrivers and drills with either a cylindrical grip or the traditional pistol grip, each available in different circumferences. Equipping a workstation with the appropriately shaped and sized handle lets workers of different strengths and hand sizes use the driver or drill. A worker using the appropriate grip is also able to keep the wrist straight and the arm close to the body to reduce the risk of overuse injury to the wrist, elbow, and shoulder.

An unintended consequence of this approach to product development is that, in many cases, the redesign is favored by groups not originally targeted. For example, modifying the entryway to a house for easy wheelchair access included eliminating steps from the garage into the house and setting the door stoop at the same level as the front walkway. This design has become popular among people with young children, because the features that make a house easily accessible to a wheelchair make it equally accessible for a stroller.

Giving a Hand to Prosthetics Design

The loss of a limb has large emotional and economic costs because of its effect on the quality of life. Biomechanics research is used to improve the design of artificial limbs (often called *prostheses* or *prosthetic limbs*). The objective is to create an artificial limb that allows as much original function as possible.

The complexity of a prosthesis depends on whether it's an arm or a leg and how much of the limb needs to be replaced. Because of the dexterity required of our fingers, a hand is probably the most difficult prosthesis to design and use. In the lower extremity, the loss of a leg below the knee means that only the ankle joint's contribution to support and locomotion is lost, while loss of the leg at mid-thigh means the contributions of both the knee and ankle joints are lost, making it challenging to coordinate the action of the two joints in a prosthetic limb.

A major focus of improving prosthetic design is to allow the user to control the action of the prosthetic limb using the muscles that remain following loss of the limb. Sensors in the socket attaching the artificial limb to the remaining natural limb can detect muscle activity. During rehabilitation, a goal is for the user to learn how to control the prosthesis using this muscle activity.

Losing Weight to Help Your Joints

Joint loading is a critical factor in the onset and progression of degenerative joint disease, or *osteoarthritis* (see Chapter 13). Heavier people are at greater risk of developing the disease, especially in the hip and knee joints. However, it isn't simply the additional weight that adds to the load but also the greater muscle force needed to move and support heavier segments.

In recent years, the Arthritis Foundation has used the statement "For every pound you gain, you add 4 pounds of pressure on your knees and six times the pressure on your hips." This statement comes from the conclusions of a clinical research project comparing the biomechanics of gait performed by normal-weight and overweight adults. The importance of the statement is that it *quantifies,* or puts a number on, the effect of greater body weight on joint loading. The statement as used by the Arthritis Foundation is intended to encourage those carrying extra weight to lose weight. Losing a single pound of body weight benefits in the prevention and progression of osteoarthritis, so any weight loss is beneficial. Biomechanics has shown that weight loss takes a large load off of the hips and knees, two joints very susceptible to osteoarthritis and that, once afflicted, hinder mobility.

Chapter 22

Ten Ways to Succeed in Your Biomechanics Course

In This Chapter

▶ Getting the most out of your time in class

▶ Using your outside-class time to your advantage

*Y*ou may have bought this book as an additional resource to help with the biomechanics class you're taking now or you will be taking soon, and for this I thank you. My years of experience teaching introductory biomechanics allow me to give you a few suggestions that may help you succeed. I don't mean just passing the class, although that's an excellent short-term goal. I mean developing an appreciation for and a foundation in biomechanics. Here are my top suggestions.

Go to Class and Ask Questions

Your mom probably told you this one. Going to class obviously lets you hear the professor's lecture and allows you to participate in the lab experiences. But there are less obvious benefits, too: Your instructor will come to recognize your face, and you'll meet other students. This adds a personal aspect to the class and allows you to more easily approach your professor, and classmates, outside of class time.

When in class, don't hesitate to ask questions — even if you're sure they're "dumb." Few students are willing to raise a hand and say, "Wait, I don't understand what you just said. Can you clarify?" Many students think they're the only ones who aren't grasping what the instructor is saying, but that's rarely the case. Lead by example and ask for clarification. Your classmates — and professor — will appreciate your initiative.

Experience has shown me that when one student doesn't understand a concept, many don't. I've seen the class perk up when one student is brave enough to ask for clarification or to pose a question.

If you aren't bold enough to speak up in class, at least write down your question and take it to the instructor after class. You may be invited to office hours if the instructor can't answer the question at that time. You may even ask the question of a classmate — together, the two of you may be able to come up with the answer, or even formulate the question in a more direct way, which you can then take to the professor's office hours.

An instructor doesn't have unlimited time during a class to show how a concept applies to all activities. The activity selected as the example may not be one that you're familiar with. Some students' immediate response is, "Well I don't do that activity, so this isn't important to me." So, think about an activity that you *are* familiar with, and consider how the concept applies to that activity instead. Ask about it in class, or bring your question to office hours.

Read the Textbook

Your instructor probably told you this one: Use your textbook as the learning resource it's intended to be.

Before the class begins a new topic, read the introduction to and summary of the appropriate chapter in your textbook. Reading *before* class gives you an overview of the new terms and concepts that you'll hear in the lecture. If you have access to slides or class notes used by your instructor, flip through those, too.

Most courses in biomechanics follow somewhat closely to the order of presentation in the course textbook, but if you're not sure which chapter or pages to read, take your textbook with you to class or office hours and ask the instructor for direction in your reading. (This is even easier if you have the electronic version of the textbook.)

After class, review your class notes. Make sure they're clear. If they don't make sense now, they won't make sense when you use them to review for an exam weeks or months down the road. Make sure any figures or diagrams you drew are correctly and completely labeled and still clear after class. Then thoroughly read the material in the chapter to get another more detailed explanation of the concepts and terms. As you read, redraw and re-label the figures in the book. Or scan the figure, cover the labels, and use this as a fill-in-the-blank review sheet.

Textbooks typically provide much more explanation than instructors can squeeze into class time. The example problems worked through in the textbook may differ from those shown during class.

Do the Problems and Review Questions at the End of the Chapter

The typical biomechanics textbook covers the concepts and shows the solutions of example problems. At the end of each chapter are additional questions. Some may be assigned as homework or as part of the hands-on portion of the course-work. Do all of them, even those not assigned.

Start by working the examples in the chapters. Check your answers and your steps to the solution. Do the same thing with examples from class — redo them and recheck them and make sure you understand all the steps involved.

 Develop a step-by-step process of problem solving that works for you. It may be the process outlined in your textbook. It may be the process outlined in this book. It may be the process your instructor uses. The best process is one that you're comfortable with, and it may include steps from different sources.

Work on each problem until you get the solution, trying it different ways. Keep your work for each attempt. If you find you're using the same approach over and over, and still not getting the solution, take a break. If you don't solve a problem after giving it a good try, go to office hours with your attempted solutions so you can show your instructor your solution process.

Create Flashcards

Purchase a set of 3-x-5-inch index cards and use them to create flashcards as new concepts or equations are introduced. Don't wait until a day or two before the exam to create the flashcards. Carry the cards with you and use them to review whenever you have some spare time. If you're more electroni-cally orientated, you can find software or a smartphone app to make elec-tronic flashcards, even some you can share with friends.

On one side of the flashcard, write down a concept definition or an equation. On the other side, write the name of the concept or the equation. For example, write *linear velocity* (explained in Chapter 5) on one side of the card, and on the other side write one version of the equation, perhaps the following:

$$v = \frac{v_f - v_i}{t_f - t_i}$$

Another card may have *linear velocity* on one side and the written definition (the rate of change of position in a specific direction) on the other side. Flashcards are cheap, so create a lot of them — for example, just for velocity, create a flashcard for each variation of the equation.

As you identify concepts and equations that give you difficulty, put these cards into a separate stack and review them more often until you learn the concept. Watching the stack of difficult concepts get smaller as you learn more is a great feeling!

Go to Office Hours

Instructors schedule regular office hours for students seeking assistance. Office hours are one of the most underused resources available to a student. The instructor is in the office for one reason: to assist you.

The busiest office hours are those on the days just before an exam. Unfortunately, many students don't realize they're having trouble until they start studying for the exam. But just before an exam is often too late to get help. You can't expect your instructor to review the entire course just for you. So, review material as the course goes along, and visit office hours regularly.

Go prepared to the office hours. Don't just walk in and say, "I don't understand." Explain what you do understand, and then point out what you find confusing and ask for clarification. Be as specific as possible. If a word problem is giving you trouble, bring your attempted solution(s) to the problem. Explain the steps you're taking to solve the problem. This gives your instructor a starting point to more easily identify why you're having trouble and give you more specific help.

The earlier you ask for help, the more help your instructor will be able, and willing, to provide.

Form a Study Group with Classmates

Arrange a regularly scheduled time with a few classmates to get together and go over the class material, lab assignments, and homework. Start meeting early in the semester. Two or more heads are better than one. Your group members can trade ideas, insights, and personal learning aids for the material. Use your group time to go over the solutions to assigned problems together. A small group is usually less intimidating than a large one, so keep the size manageable.

Accept and Apply Newton as the Foundation of Movement Analysis

There are very few precise cause-and-effect relationships in the universe. One is Newton's first law: An unbalanced force causes a change in motion. Accept the law as a 300-year-old truth (Newton first proposed his laws in 1687), and apply it to any movement you're watching.

When learning a new movement, or practicing an old one, identify all the forces involved. Look at the movement of the body in terms of its momentum: how fast and in what direction it's moving. Most bodies move in more than one direction at a time, and the motion can change in each direction. To change momentum, an unbalanced force must act on the body for a period of time. Question what you currently know, as well as explanations others give you, about how things move and change motion. Base your questioning on Newton's laws.

When provided with a checklist of points describing how a skill is performed or problems associated with a disease or disability, look for the biomechanics concept that underlies each point. Ask yourself if the listed point is an error, a symptom of an error, or an idiosyncrasy (see Chapter 16). Many checklists include points that are idiosyncrasies of performance or a symptom of an error. Focusing attention on these points won't improve performance because the cause — a violation of a biomechanical principle or a modification of performance by the individual to account for a lacking critical ability — is not addressed.

Talk Fluent Biomechanics with Your Classmates

Don't limit your learning about biomechanics to the classroom, the lab, your desk, or your study group. Using the terms and applying the concepts of biomechanics when analyzing any movement is a good way to reinforce the ideas.

When watching or discussing a live or televised event with friends, challenge each other to apply a principle of biomechanics to a skill used during the event. Compare the styles of different pitchers in a baseball game, looking for similarities and differences. What principles of biomechanics apply in a diving competition? How do dancers, figure skaters, or gymnasts rotate, twist, and spin? How is the elegant form of a dancer performing a grande jette similar to or different from a power forward in basketball performing a slam dunk?

Learn from each other by discussing how the ideas covered in your biomechanics class relate to the different activities each group member finds interesting. Combine your existing understanding of how a particular skill is performed with the principles of biomechanics. We all have existing knowledge of how things move — our own explanations of how movement is performed successfully. Question the accuracy of this knowledge. Reconcile your preexisting ideas with the principles of biomechanics.

You may have a mental checklist of what points to look for in performance of a particular skill, so take each of the points in the checklist and see if there is an underlying principle to justify it. Most people have some experience with some form of skilled movement. Many have reached a high level of proficiency in performance, and they think they just "do it." But just because you can perform a skill doesn't mean you understand it or that you can teach others to do it. Students often don't even recognize this until they're put in a position of explaining and teaching a familiar skill as a coach or instructor.

In general, be curious about moving things and look for the principles of biomechanics common across them all.

Volunteer for Research Projects

 Ask your instructor if there are opportunities for you to get involved in any research projects going on in the biomechanics lab. Depending on the type of research conducted, you may be able to serve as a subject if you meet the selection criteria. Criteria may include a certain level of experience with the task being evaluated, a particular prior injury history, or being within a specific age range. If you get the opportunity to serve as a subject, arrive before the scheduled time. Make sure you have with you what you were asked to bring, whether it's specific clothing, equipment, or additional information sheets.

 If you can't participate in any research projects, ask if you can observe the data collection process. Interested and dependable students are almost always welcome to find out more about the project. If you get the opportunity to observe, do a little preparation beforehand. Ask your instructor for a copy of the consent form that will be signed by the research participants. The consent form provides an overview of the study, including the purpose of the project, an explanation of how the data will be collected, and the expected results. When you go to observe, take along a notebook and write down questions or comments that come up as you watch what's going on. Finally, show up on time — you don't want to disrupt the project.

Don't ask questions while the data collection is underway. Save questions and comments for after the data collection is complete and the subject has left the lab. The investigator, time allowing, will be willing to answer your questions and discuss your comments.

Attend a Biomechanics Conference

Most professional organizations, including those in biomechanics, hold regular annual meetings attended by members of the organization. Conferences provide a forum for disseminating and discussing research and knowledge in biomechanics. For students, conferences present an opportunity for learning more about biomechanics and for networking with established researchers and other students.

Start with a student-focused conference, like a regional meeting hosted by the American Society of Biomechanics (ASB) or the American College of Sports Medicine (ACSM). The registration fee for students at these conferences is kept low to encourage attendance.

A conference typically includes a variety of different presentation formats. The highlight is often a keynote presentation of 40 to 60 minutes, given by a speaker who was invited because of an outstanding record of accomplishment in biomechanics.

Shorter presentations by individual speakers representing a research group are also scheduled, in podium and poster sessions. A podium session includes four to eight individual 15-minute presentations made orally. The speaker presents an overview of a research project for 10 to 12 minutes, followed by questions from the audience. A poster session includes multiple presenters who set up a poster of a research project and then interact with attendees who come by to discuss the poster.

Many conferences also include additional sessions dedicated to topics that could include applying for graduate school, data processing, or a current hot topic in biomechanics.

The schedule for a conference is usually posted on the website at least one week before the start date. Look over the schedule early to identify any special sessions you want to be sure to attend. The schedule will also include breaks and other opportunities for informal conversations related to biomechanics and networking. Attending a conference provides a great chance to expand your understanding of biomechanics and to meet others interested in the field.

Index

● *C* ●

• *N* •

Notes

Notes

Notes

Notes

Notes

Notes

About the Author

Steven T. McCaw, PhD, has been on the faculty of the School of Kinesiology and Recreation at Illinois State University since 1989. His instructional responsibilities include Biomechanics of Human Movement, Quantitative Analysis in Biomechanics, Occupational Biomechanics, and Advanced Biomechanics. He has served as chair of over 40 master's theses and has published over 40 manuscripts in refereed journals. Steve received his PhD in Biomechanics/Sports Medicine from the University of Oregon in Eugene; received his Master of Arts from McGill University in Montreal, Canada; and completed his undergraduate work at Lakehead University, Thunder Bay, Canada.

Dedication

To my wife, Kathy, the constant support of this journey.

Author's Acknowledgments

To the supportive staff at John Wiley & Sons, especially my acquisitions editor, Anam Ahmed, and my project editor, Elizabeth Kuball, for their guidance in the development and completion of this book. Dr. Monique Mokha's comments as the technical editor are deeply appreciated.

Publisher's Acknowledgments

Acquisitions Editor: Anam Ahmed

Project Editor: Elizabeth Kuball

Copy Editor: Elizabeth Kuball

Technical Editor: Monique Mokha, PhD

Art Coordinator: Alicia B. South

Project Coordinator: Patrick Redmond

Illustrator: Kathryn Born

Cover Image: © iStockphoto.com/ John_Woodcock

Apple & Mac

iPad For Dummies,
5th Edition
978-1-118-49823-1

iPhone 5 For Dummies,
5th Edition
978-1-118-35201-4

MacBook For Dummies,
4th Edition
978-1-118-20920-2

OS X Mountain Lion
For Dummies
978-1-118-39418-2

Blogging & Social Media

Facebook For Dummies,
5th Edition
978-1-118-09562-1

Mom Blogging
For Dummies
978-1-118-03843-7

Pinterest For Dummies
978-1-118-32800-2

WordPress For Dummies,
5th Edition
978-1-118-38318-6

Business

Commodities For Dummies,
2nd Edition
978-1-118-01687-9

Investing For Dummies,
6th Edition
978-0-470-90545-6

Personal Finance
For Dummies,
7th Edition
978-1-118-11785-9

QuickBooks 2013
For Dummies
978-1-118-35641-8

Small Business Marketing Kit
For Dummies,
3rd Edition
978-1-118-31183-7

Careers

Job Interviews
For Dummies,
4th Edition
978-1-118-11290-8

Job Searching with
Social Media
For Dummies
978-0-470-93072-4

Personal Branding
For Dummies
978-1-118-11792-7

Resumes For Dummies,
6th Edition
978-0-470-87361-8

Success as a Mediator
For Dummies
978-1-118-07862-4

Diet & Nutrition

Belly Fat Diet For Dummies
978-1-118-34585-6

Eating Clean For Dummies
978-1-118-00013-7

Nutrition For Dummies,
5th Edition
978-0-470-93231-5

Digital Photography

Digital Photography
For Dummies,
7th Edition
978-1-118-09203-3

Digital SLR Cameras &
Photography For Dummies,
4th Edition
978-1-118-14489-3

Photoshop Elements 11
For Dummies
978-1-118-40821-6

Gardening

Herb Gardening
For Dummies,
2nd Edition
978-0-470-61778-6

Vegetable Gardening
For Dummies,
2nd Edition
978-0-470-49870-5

Health

Anti-Inflammation Diet
For Dummies
978-1-118-02381-5

Diabetes For Dummies,
3rd Edition
978-0-470-27086-8

Living Paleo For Dummies
978-1-118-29405-5

Hobbies

Beekeeping
For Dummies
978-0-470-43065-1

eBay For Dummies,
7th Edition
978-1-118-09806-6

Raising Chickens
For Dummies
978-0-470-46544-8

Wine For Dummies,
5th Edition
978-1-118-28872-6

Writing Young Adult Fiction
For Dummies
978-0-470-94954-2

Language &
Foreign Language

500 Spanish Verbs
For Dummies
978-1-118-02382-2

English Grammar
For Dummies,
2nd Edition
978-0-470-54664-2

French All-in One
For Dummies
978-1-118-22815-9

German Essentials
For Dummies
978-1-118-18422-6

Italian For Dummies
2nd Edition
978-1-118-00465-4

 **Available in print and e-book formats.**

Math & Science

Algebra I For Dummies,
2nd Edition
978-0-470-55964-2

Anatomy and Physiology
For Dummies,
2nd Edition
978-0-470-92326-9

Astronomy For Dummies,
3rd Edition
978-1-118-37697-3

Biology For Dummies,
2nd Edition
978-0-470-59875-7

Chemistry For Dummies,
2nd Edition
978-1-1180-0730-3

Pre-Algebra Essentials
For Dummies
978-0-470-61838-7

Microsoft Office

Excel 2013 For Dummies
978-1-118-51012-4

Office 2013 All-in-One
For Dummies
978-1-118-51636-2

PowerPoint 2013
For Dummies
978-1-118-50253-2

Word 2013 For Dummies
978-1-118-49123-2

Music

Blues Harmonica
For Dummies
978-1-118-25269-7

Guitar For Dummies,
3rd Edition
978-1-118-11554-1

iPod & iTunes
For Dummies,
10th Edition
978-1-118-50864-0

Programming

Android Application
Development For
Dummies, 2nd Edition
978-1-118-38710-8

iOS 6 Application
Development For Dummies
978-1-118-50880-0

Java For Dummies,
5th Edition
978-0-470-37173-2

Religion & Inspiration

The Bible For Dummies
978-0-7645-5296-0

Buddhism For Dummies,
2nd Edition
978-1-118-02379-2

Catholicism For Dummies,
2nd Edition
978-1-118-07778-8

Self-Help & Relationships

Bipolar Disorder
For Dummies,
2nd Edition
978-1-118-33882-7

Meditation For Dummies,
3rd Edition
978-1-118-29144-3

Seniors

Computers For Seniors
For Dummies,
3rd Edition
978-1-118-11553-4

iPad For Seniors
For Dummies,
5th Edition
978-1-118-49708-1

Social Security
For Dummies
978-1-118-20573-0

Smartphones & Tablets

Android Phones
For Dummies
978-1-118-16952-0

Kindle Fire HD
For Dummies
978-1-118-42223-6

NOOK HD For Dummies,
Portable Edition
978-1-118-39498-4

Surface For Dummies
978-1-118-49634-3

Test Prep

ACT For Dummies,
5th Edition
978-1-118-01259-8

ASVAB For Dummies,
3rd Edition
978-0-470-63760-9

GRE For Dummies,
7th Edition
978-0-470-88921-3

Officer Candidate Tests,
For Dummies
978-0-470-59876-4

Physician's Assistant Exam
For Dummies
978-1-118-11556-5

Series 7 Exam
For Dummies
978-0-470-09932-2

Windows 8

Windows 8 For Dummies
978-1-118-13461-0

Windows 8 For Dummies,
Book + DVD Bundle
978-1-118-27167-4

Windows 8 All-in-One
For Dummies
978-1-118-11920-4

Available in print and e-book formats.

Take Dummies with you everywhere you go!

Whether you're excited about e-books, want more from the web, must have your mobile apps, or swept up in social media, Dummies makes everything easier .

Dummies products make life easier

- DIY
- Consumer Electronics
- Crafts
- Software
- Cookware
- Hobbies
- Videos
- Music
- Games
- and More!

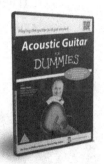

For more information, go to **Dummies.com**® and search the store by category.

FOR DUMMIES
A Wiley Brand